Lecture Notes in Statistics

Edited by J. Berger, S. Fienberg, J. Gani, K. Krickeberg, I. Olkin, and B. Singer

59

Sudhir Gupta
Rahul Mukerjee

A Calculus for Factorial Arrangements

Springer-Verlag Berlin Heidelberg GmbH

Authors

Sudhir Gupta
Division of Statistics, Department of Mathematical Sciences
Northern Illinois University
DeKalb, IL 60115, USA

Rahul Mukerjee
Stat-Math Division, Indian Statistical Institute
203, B.T. Road, Calcutta 700 035, India

Mathematical Subject Classification: 62K15, 62K10

ISBN 978-0-387-97172-8 ISBN 978-1-4419-8730-3 (eBook)
DOI 10.1007/978-1-4419-8730-3

Originally published by Springer-Verlag Berlin Heidelberg in 1989

2847/3140-543210 – Printed on acid-free paper

TO

Suman, Piyush, Gaorav
SG

My parents and wife
RM

ACKNOWLEDGEMENTS

Most parts of this monograph were written during the 1987-88 academic year when one of the authors (RM) was visiting the Department of Mathematical Sciences, Northern Illinois University. We are thankful to the department and our colleagues for a very friendly and congenial atmosphere during the preparation of this monograph. We also thank Professors W. T. Federer and J. N. Srivastava for some helpful comments.

The first author (SG) is grateful to his wife Suman and sons Piyush and Gaorav for their love, support and affection throughout. The second author (RM) wishes to record his deep gratitude to Professor S. K. Chatterjee, Department of Statistics, University of Calcutta, who had inspired and initiated him to this area of research. He is also grateful to his parents and wife for their love and support in various ways.

Finally we thank Mrs. Sara Clayton for her diligent typing of the monograph and Ms. Susan Gordon of Springer-Verlag Inc. for her help at various stages.

Sudhir Gupta
Rahul Mukerjee

Table of Contents

CHAPTER 1

INTRODUCTION

Factorial designs were introduced and popularized by Fisher (1935). Among the early authors, Yates (1937) considered both symmetric and asymmetric factorial designs. Bose and Kishen (1940) and Bose (1947) developed a mathematical theory for symmetric prime-powered factorials while Nair and Rao (1941, 1942, 1948) introduced and explored balanced confounded designs for the asymmetric case. Since then, over the last four decades, there has been a rapid growth of research in factorial designs and a considerable interest is still continuing.

Kurkjian and Zelen (1962, 1963) introduced a tensor calculus for factorial arrangements which, as pointed out by Federer (1980), represents a powerful statistical analytic tool in the context of factorial designs. Kurkjian and Zelen (1963) gave the analysis of block designs using the calculus and Zelen and Federer (1964) applied it to the analysis of designs with two-way elimination of heterogeneity. Zelen and Federer (1965) used the calculus for the analysis of designs having several classifications with unequal replications, no empty cells and with all the interactions present. Federer and Zelen (1966) considered applications of the calculus for factorial experiments when the treatments are not all equally replicated, and Paik and Federer (1974) provided extensions to when some of the treatment combinations are not included in the experiment.

The calculus, which involves the use of Kronecker products of matrices, is extremely helpful in deriving characterizations, in a compact form, for various important features like balance and orthogonality in a general multifactor setting. In turn, these characterizations lead to useful general construction procedures for designs with desirable properties. Recently, it has been seen that the calculus is helpful even in possibly non-orthogonal situations. Furthermore, as recent results reveal, the calculus enables a unification of the classical construction procedures. The fact that a major part of the work on (complete) factorial designs since the 1970s has been in or around the calculus further demonstrates its importance.

This monograph aims at presenting a survey of these developments in or relating to the calculus since the 1970s. The basic notions and preliminaries for the calculus have been introduced in Chapter 2. In Chapters 3 and 4, it has been seen how the calculus serves as a very powerful tool in the derivation of characterizations for balance and orthogonality in factorial designs. These characterizations have immediate applicability and yield useful methods of construction with wide coverage as discussed in Chapters 5 and 6. The direct and indirect applications of the calculus in integrating the classical construction procedures have been reviewed in Chapter 7. Chapter 8 presents some other developments including some more methods of construction, results on efficiency and several other applications of the calculus.

The applications of the calculus, with reference to complete factorials, have been considered in this monograph. The theory and methods of fractional replication have been left out of consideration since excellent accounts of the developments in that area up to various stages have already been given by Srivastava (1978), Raktoe, Hedayat and Federer (1980) and Dey (1985). For further references on fractional replication, including search designs, the interested reader may see Srivastava (1987), Kuwada and Nishii (1988) and Chatterjee and Mukerjee (1986).

It has been attempted to make the monograph self-contained. However, some prior knowledge in the general area of experimental design will be helpful and, in this context, reference is made to Kempthorne (1952), Federer (1955), John (1971), Raghavarao (1971) or Street and Street (1987). For the relevant linear estimation theory, we refer to Rao (1973a). Notational uniformity has been retained within each chapter. However, some overlapping of notation between chapters has been unavoidable. For example, a treatment combination has been denoted by both $(j_1, \ldots, j_n)$ and $j_1 \cdots j_n$ depending on the notational convenience. Nevertheless, it is hoped that the notational system will be clear from the context.

CHAPTER 2
A CALCULUS FOR FACTORIAL ARRANGEMENTS

2.1. Introduction

Consider a factorial experiment involving n factors, $F_1, F_2, \ldots, F_n$, at $m_1, m_2, \ldots, m_n$ (≥ 2) levels respectively. Let the levels of F_i be coded as $0, 1, \ldots, m_i - 1$ $(1 \leq i \leq n)$. A typical selection of levels $j = (j_1, j_2, \ldots, j_n)$, $0 \leq j_i \leq m_i - 1$, $1 \leq i \leq n$, will be termed the jth treatment combination and the effect due to this treatment combination will be denoted by $\tau(j_1, j_2, \ldots, j_n)$. There are $v = \prod_{i=1}^{n} m_i$ treatment combinations in all.

The treatment effects, namely $\tau(j_1, j_2, \ldots, j_n)'s$ are unknown parameters. A linear parametric function

$$\sum_{j_1} \sum_{j_2} \cdots \sum_{j_n} l_{j_1 j_2 \cdots j_n} \tau(j_1, j_2, \ldots, j_n), \tag{2.1.1}$$

where $\{l_{j_1 j_2 \cdots j_n}\}$ are real numbers, not all zeros, such that $\sum_{j_1} \sum_{j_2} \cdots \sum_{j_n} l_{j_1 j_2 \cdots j_n} = 0$, will be called a treatment contrast. In a factorial experiment, interest lies in special types of contrasts - namely those belonging to the interactions. Specifically, following Bose (1947), a treatment contrast of the form (2.1.1) will be said to belong to the g-factor interaction $F_{i_1} F_{i_2} \cdots F_{i_g} (1 \leq i_1 < i_2 < \cdots < i_g \leq n, 1 \leq g \leq n)$ if $l_{j_1 j_2 \cdots j_n}$ depends only on $j_{i_1}, j_{i_2}, \ldots, j_{i_g}$ and (ii) the sum of $l_{j_1 j_2 \cdots j_n}$ over any one of the arguments $j_{i_1}, j_{i_2}, \ldots, j_{i_g}$ is zero. Because of (ii), it follows that there are $\prod_{u=1}^{g} (m_{i_u} - 1)$ linearly independent treatment contrasts belonging to the interaction $F_{i_1} F_{i_2} \cdots F_{i_g}$. The 1-factor interactions (i.e., $g = 1$) will be called main effects. Obviously, there are $2^n - 1$ interactions in all.

Example 2.1.1. Let $n = 2$. Then a typical contrast belonging to main effect F_1 is of the form $\sum_{j_1} l_{j_1} \tau(j_1, j_2)$, where $\{l_{j_1}\}$ are real numbers, not all zeros, satisfying

$\sum_{j_1} l_{j_1} = 0$. Similarly, one may consider a contrast belonging to main effect F_2. Incidentally, the same notation is being used for a factor and the corresponding main effect. A typical contrast belonging to the 2-factor interaction F_1F_2 is given by $\sum_{j_1}\sum_{j_2} l_{j_1j_2}\tau(j_1,j_2)$, where $\{l_{j_1j_2}\}$ are real numbers, not all zeros, satisfying $\sum_{j_1} l_{j_1j_2} = 0$, $\forall j_2$; $\sum_{j_2} l_{j_1j_2} = 0$, $\forall j_1$.

2.2. Elements and operations of the calculus

The calculus for factorial arrangements provides a very powerful tool for expressing the notation in a very compact and convenient form. The calculus was introduced by Kurkjian and Zelen (1962, 1963) although it appears that some of their ideas were also inherent in Zelen (1958) and Shah (1958).

Let $\underline{a}_i = (0,1,\dots,m_i-1)'$, $1 \le i \le n$. Throughout this monograph, the v treatment combinations will be considered in the lexicographic order given by $\underline{a}_1 \times \underline{a}_2 \times \cdots \times \underline{a}_n$, where $\times$ denotes symbolic direct product as defined by Shah (1958). Let $\underline{\tau}$ be a $v \times 1$ vector, with elements given by the $\tau(j_1,j_2,\dots,j_n)'s$ arranged lexicographically. For example, if $n = 2$, $m_1 = 2$, $m_2 = 3$, then $\underline{a}_1 = (0,1)'$, $\underline{a}_2 = (0,1,2)'$, $\underline{a}_1 \times \underline{a}_2 = (00,01,02,10,11,12)'$ and $\underline{\tau} = (\tau(0,0),\tau(0,1),\tau(0,2), \tau(1,0),\tau(1,1),\tau(1,2))'$.

By (2.1.1), a typical treatment contrast is of the form $\underline{l}'\,\underline{\tau}$, where the $v \times 1$ coefficient vector $\underline{l}$ is non-null and the sum of the elements of $\underline{l}$ equals zero. Such a contrast will be said to be normalized if $\underline{l}'\,\underline{l} = 1$. Two treatment contrasts $\underline{l}_1'\,\underline{\tau}$ and $\underline{l}_2'\,\underline{\tau}$ will be called mutually orthogonal if $\underline{l}_1'\,\underline{l}_2 = 0$. A set of treatment contrasts will be called orthonormal if the contrasts in the set are all normalized and mutually orthogonal.

Let Ω be the set of all n-component non-null binary vectors. It is easy to see that there is a one-one-correspondence between Ω and the set of all interactions, in the sense that a typical interaction $F_{i_1}F_{i_2}\cdots F_{i_g}$ $(1 \le i_1 < i_2 < \cdots < i_g \le n,\ 1 \le g \le n)$ corresponds to the element $x = (x_1,\dots,x_n)$ of Ω such that $x_{i_1} = x_{i_2} = \cdots = x_{i_g} = 1$ and $x_u = 0$ for $u \ne i_1, i_2,\dots,i_g$. Thus the $2^n - 1$ interactions may be denoted by F^x, $x \in \Omega$. For example, if $n = 2$, then the main effects F_1, F_2 and the 2-factor interaction F_1F_2 may be denoted by F^{10}, F^{01} and F^{11} respectively. The treatment contrasts belonging to the interactions may be conveniently represented making use of Kronecker products, as indicated below.

For each $x = (x_1, \ldots, x_n) \in \Omega$, let

$$M^x = M_1^{x_1} \otimes M_2^{x_2} \otimes \cdots \otimes M_n^{x_n} = \bigotimes_{i=1}^{n} M_i^{x_i}, \tag{2.2.1}$$

where $\otimes$ denotes Kronecker product and for $1 \le i \le n$,

$$\left.\begin{aligned} M_i^{x_i} &= I_i - m_i^{-1} J_i \qquad \text{if } x_i = 1 \\ &= m_i^{-1} J_i \qquad\qquad \text{if } x_i = 0 \end{aligned}\right\} \tag{2.2.2}$$

Here I_i is an identity matrix and J_i is a matrix of 1's both of order $m_i \times m_i$.

Lemma 2.2.1. For each $x \in \Omega$, the elements of $M^x \underline{\tau}$ represent a complete set of treatment contrasts belonging to the interaction F^x.

Proof. It is easy to see that the elements of $M^x \underline{\tau}$ are treatment contrasts belonging to F^x. The proof may now be completed by observing that in view of (2.2.1), (2.2.2),

$$rank(M^x) = \prod_{i=1}^{n} rank(M_i^{x_i}) = \prod_{i=1}^{n} (m_i - 1)^{x_i},$$

which is the same as the maximum number of linearly independent contrasts belonging to F^x.

Lemma 2.2.2. Treatment contrasts belonging to any two distinct interactions are mutually orthogonal.

Proof. By Lemma 2.2.1, it is enough to show that for every $x = (x_1, \ldots, x_n)$ and $y = (y_1, \ldots, y_n) \in \Omega$, $x \neq y$,

$$M^x M^y = 0. \tag{2.2.3}$$

Now, by (2.2.1) and the standard rules for operations with Kronecker products,

$$M^x M^y = \bigotimes_{i=1}^{n} (M_i^{x_i} M_i^{y_i}). \tag{2.2.4}$$

Now, $x \neq y$ implies that $x_i \neq y_i$ for at least one i, and for this i, by (2.2.2.), $M_i^{x_i} M_i^{y_i} = 0$. Hence (2.2.3) follows from (2.2.4).

Lemma 2.2.1 gives a representation for the treatment contrasts belonging to the different interactions. Another equivalent representation in terms of orthonormal contrasts is often helpful. For $1 \leq i \leq n$, let 1_i be the $m_i \times 1$ vector with all elements unity and P_i be an $(m_i - 1) \times m_i$ matrix such that the $m_i \times m_i$ matrix $(m_i^{-1/2} 1_i, P_i')'$ is orthogonal. For example, if $n = 2$, $m_1 = 2, m_2 = 3$, then one may take

$$P_1 = \begin{pmatrix} \frac{1}{\sqrt{2}} & -\frac{1}{\sqrt{2}} \end{pmatrix}, \quad P_2 = \begin{pmatrix} \frac{1}{\sqrt{2}} & -\frac{1}{\sqrt{2}} & 0 \\ \frac{1}{\sqrt{6}} & \frac{1}{\sqrt{6}} & -\frac{2}{\sqrt{6}} \end{pmatrix}$$

For each $x = (x_1, \ldots, x_n) \in \Omega$, let

$$P^x = \bigotimes_{i=1}^{n} P_i^{x_i}, \tag{2.2.5}$$

where

$$\left.\begin{aligned} P_i^{x_i} &= P_i && \text{if } x_i = 1 \\ &= m_i^{-1/2}\, 1_i' && \text{if } x_i = 0 \end{aligned}\right\} \quad (1 \leq i \leq n). \tag{2.2.6}$$

By (2.2.2) and (2.2.6), the relation $P_i^{x_i\prime} P_i^{x_i} = M_i^{x_i}$ holds for every i, whether x_i equals 0 or 1. Hence by (2.2.1) and (2.2.5), for every $x \in \Omega$,

$$P^{x\prime} P^x = M^x. \tag{2.2.7}$$

Also, analogously to (2.2.3), it may be seen that for each $x, y \in \Omega$, $x \neq y$,

$$P^x P^{x'} = I, \quad P^x P^{y'} = 0, \tag{2.2.8}$$

where I is an identity matrix of appropriate order. Hence our next lemma follows along the line of Lemma 2.2.1.

Lemma 2.2.3. For each $x \in \Omega$, the elements of $P^x \underline{\tau}$ represent a complete set of orthonormal contrasts belonging to the interaction F^x.

In the sequel, both the representations, as given in Lemmas 2.2.1 and 2.2.3 above, will be found to be useful.

2.3. Orthogonal factorial structure and balance

In this section, two fundamental concepts in the context of factorial experiments will be introduced. This will be done primarily with reference to a block design.

Consider an arrangement of the $v = \Pi m_i$ treatment combinations in a block design involving b blocks of sizes $k_1, k_2, \ldots, k_b$, the jth treatment combination being replicated r_j times. The design will be called proper if $k_1, k_2, \ldots, k_b$ are all equal and equireplicate if the $r_j's$ are all equal. The $v \times b$ matrix $N = ((n_{jh}))$ will be termed the incidence matrix of the design where $n_{jh} (\geq 0)$ is the number of times the jth treatment combination occurs in the hth block. Let $\underline{r} = (r_1, r_2, \ldots, r_v)'$, $\underline{k} = (k_1, k_2, \ldots, k_b)'$, $r^\delta = Diag(r_1, r_2, \ldots, r_v)$, $k^\delta = Diag(k_1, k_2, \ldots, k_b)$. The fixed effects intrablock model with independent homoscedastic errors (the constant error variance being, say, σ^2) and no block versus treatment interaction will be assumed. Then it is well-known (see e.g., Raghavarao (1971)) that the intrablock reduced normal equations for the vector of treatment effects $\underline{\tau}$ are given by

$$C\underline{\tau} = \underline{Q} \tag{2.3.1}$$

where

$$C = r^\delta - Nk^{-\delta}N', \tag{2.3.2}$$

is the usual C-matrix of the design, $k^{-\delta} = (k^\delta)^{-1}$ and $\underline{Q}$ is the vector of adjusted treatment totals.

From (2.3.2), C is a symmetric matrix with all row sums zero. Hence $rank\ (C) \leq v - 1$. A design is called connected if $rank\ (C) = v - 1$. A treatment contrast $\underline{l}'\ \underline{\tau}$ is estimable if $\underline{l}' \in R(C)$, where for any matrix A, $R(A)$ stands for its row space. Clearly, for an estimable treatment contrast $\underline{l}'\ \underline{\tau}$, there exists a $v \times 1$ vector $\underline{l}^*$ such that $\underline{l}' = \underline{l}^{*\prime} C$, and it may be seen that the best linear unbiased estimator (BLUE) of $\underline{l}'\ \underline{\tau}$ is given by $\underline{l}'\hat{\underline{\tau}} = \underline{l}^{*\prime} \underline{Q}$. It is also well-known that all treatment contrasts are estimable if and only if the design is connected.

In analyzing the results of a factorial design, the experimenter is primarily interested in drawing conclusion on the contrasts belonging to the different interactions. A great simplification occurs in interpreting the results of analysis if the design has orthogonal factorial structure in the sense of Definition 2.3.1 given below. Another important usefulness of OFS is realized while constructing confidence intervals for estimable contrasts belonging to different interactions. A discussion on the importance of orthogonal estimation of parameters in constructing confidence intervals can be found in Box and Draper (1987).

Definition 2.3.1. A factorial design will be said to have the orthogonal factorial structure (OFS) if the BLUE's of estimable treatment contrasts belonging to distinct interactions are mutually orthogonal, i.e., uncorrelated.

In other words, by Lemmas 2.2.1 and 2.2.3, OFS holds if for each $x, y \in \Omega$, $x \neq y$, the BLUE of every estimable linear combination of the elements of $M^x \underline{\tau}$ (or $P^x \underline{\tau}$) is uncorrelated with the BLUE of every estimable linear combination of the elements of $M^y \underline{\tau}$ (or $P^y \underline{\tau}$). When this is realized, in the connected case, the adjusted treatment sum of squares (SS) can be split up orthogonally into components due to the different interactions and, as such, these components may be shown in the same analysis of variance (ANOVA) table. The same can be done also in the disconnected case provided some further conditions, as elaborated in the subsequent chapters, hold. Incidentally, in Lemma 2.2.2, it was shown that contrasts belonging to distinct interactions are mutually orthogonal. The notion of OFS calls for a reflection of this property also in terms of the BLUE's of such contrasts.

Another important and useful concept in the context of factorial designs is that of balance. A definition of balance, along the line of Shah (1958), is as follows.

Definition 2.3.2. In a factorial design, an interaction F^x, $x \in \Omega$, will be said to be balanced if either (a) all treatment contrasts belonging to F^x are estimable and the BLUE's of all normalized contrasts belonging to F^x have the same variance or (b) no contrast belonging to F^x is estimable.

A factorial design will be called balanced if F^x is balanced for every $x \in \Omega$. In Definition 2.3.2, the trivial situation (b) has been included mainly for mathematical completeness; this situation will never arise if, in particular, the design is connected. The following lemma provides an interpretation for balance which is useful in practice.

Lemma 2.3.1. In a factorial design, an interaction F^x is balanced in the sense (a) of Definition 2.3.2 if and only if all treatment contrasts belonging to F^x are estimable and the BLUE's of every two mutually orthogonal contrasts belonging to F^x are uncorrelated.

Proof.

Only if: Suppose the interaction F^x is balanced in the sense (a) of Definition 2.3.2. Let $\underline{l}_1'\underline{\tau}$ and $\underline{l}_2'\underline{\tau}$ be two mutually orthogonal treatment contrasts belonging to F^x. Define $\underline{\xi}_i = (\underline{l}_i'\underline{l}_i)^{-1/2}\underline{l}_i (i = 1,2)$, and $\underline{\xi} = \frac{1}{\sqrt{2}}(\underline{\xi}_1+\underline{\xi}_2)$, and note that $\underline{\xi}_1'\underline{\tau}$, $\underline{\xi}_2'\underline{\tau}$ and $\underline{\xi}'\underline{\tau}$ are normalized contrasts belonging to F^x. Since F^x is balanced in the sense (a) of Definition 2.3.2,

$$Var(\underline{\xi}'\hat{\underline{\tau}}) = Var(\underline{\xi}_1'\hat{\underline{\tau}}) = Var(\underline{\xi}_2'\hat{\underline{\tau}}). \tag{2.3.3}$$

Again,

$$Var(\underline{\xi}'\hat{\underline{\tau}}) = \frac{1}{2}\,Var(\underline{\xi}_1'\hat{\underline{\tau}} + \underline{\xi}_2'\hat{\underline{\tau}}) = \frac{1}{2}\,\{Var(\underline{\xi}_1'\hat{\underline{\tau}}) + Var(\underline{\xi}_2'\hat{\underline{\tau}}) + 2\,Cov(\underline{\xi}_1'\hat{\underline{\tau}},\underline{\xi}_2'\hat{\underline{\tau}})\},$$

which, together with (2.3.3), yields $Cov(\underline{\xi}_1'\hat{\underline{\tau}},\underline{\xi}_2'\hat{\underline{\tau}}) = 0$ and hence, $Cov(\underline{l}_1'\hat{\underline{\tau}},\underline{l}_2'\hat{\underline{\tau}}) = 0$, as desired.

If: Let the condition stated in the lemma hold. Consider any two distinct normalized contrasts $\underline{\xi}_1'\underline{\tau}$ and $\underline{\xi}_2'\underline{\tau}$ belonging to F^x. If $\underline{\xi}_1 = -\underline{\xi}_2$, then trivially $Var(\underline{\xi}_1'\hat{\underline{\tau}}) = Var(\underline{\xi}_2'\hat{\underline{\tau}})$. Otherwise, $(\underline{\xi}_1+\underline{\xi}_2)'\underline{\tau}$ and $(\underline{\xi}_1-\underline{\xi}_2)'\underline{\tau}$ are mutually orthogonal treatment contrasts belonging to F^x, and hence under the condition stated in the lemma, $Cov\{(\underline{\xi}_1+\underline{\xi}_2)'\hat{\underline{\tau}},\ (\underline{\xi}_1-\underline{\xi}_2)'\hat{\underline{\tau}}\} = 0$, which yields $Var(\underline{\xi}_1'\hat{\underline{\tau}}) = Var(\underline{\xi}_2'\hat{\underline{\tau}})$.

The following corollary is an immediate consequence of Lemmas 2.2.3 and 2.3.1.

Corollary 2.3.1. In a factorial design, an interaction F^x is balanced in the sense (a) of Definition 2.3.2. if and only if all treatment contrasts belonging to F^x are estimable and the dispersion matrix of $P^x\hat{\underline{\tau}}$ is proportional to the identity matrix.

While OFS ensures 'between-interaction' orthogonality, by Lemma 2.3.1 balance ensures 'within-interaction' orthogonality. Therefore, if a design has OFS and is balanced then further simplifications in interpreting confidence intervals and results of the analysis are achieved. Furthermore, as we shall see later on in Chapter 8, such designs have elegant properties from efficiency considerations as well. It is, therefore, of interest to explore the algebraic and combinatorial characterizations for designs which are balanced with OFS. This has been taken up in the next chapter. The results will be seen to be useful in the actual construction of designs.

While concluding this section, it may be remarked that all the definitions in this section, as well as Lemma 2.3.1, may be easily extended to row-column designs and designs for multiway elimination of heterogeneity by interpreting C as the coefficient matrix of the reduced normal equations for treatment effects in such situations.

CHAPTER 3
CHARACTERIZATIONS FOR BALANCE WITH ORTHOGONAL FACTORIAL STRUCTURE

3.1 Algebraic characterizations

This chapter considers factorial designs which are balanced and have orthogonal factorial structure (OFS). Such designs have been termed balanced factorial experiments by Shah (1958, 1960a). They are also known as balanced confounded designs according to the nomenclature of Nair and Rao (1948). The main result of this section, namely Theorem 3.1.1, gives an algebraic characterization for balance with OFS in the connected case. For equireplicate and proper designs, the 'sufficiency' part of this result was proved by Kurkjian and Zelen (1963), while the 'necessity' part was proved by Kshirsagar (1966). Gupta (1983a) considered extensions to designs that are not necessarily equireplicate or proper. The following definition and lemmas will be helpful.

Let Ω^* be the set of all n-component binary vectors, i.e., $\Omega^* = \Omega \cup \{(0,0,...,0)\}$, where Ω is as defined in Chapter 2. For $x = (x_1, \ldots, x_n) \in \Omega^*$, let

$$Z^x = \bigotimes_{i=1}^{n} Z_i^{x_i}, \tag{3.1.1}$$

where for $1 \leq i \leq n$,

$$\left.\begin{aligned} Z_i^{x_i} &= I_i \quad \text{if } x_i = 1 \\ &= J_i \quad \text{if } x_i = 0 \end{aligned}\right\}. \tag{3.1.2}$$

Definition 3.1.1. A $v \times v$ matrix G, where $v = \Pi m_i$, will be said to have property A if it is of the form

$$G = \sum_{x \in \Omega^*} h(x) Z^x,$$

where $h(x)$, $x \in \Omega^*$, are real numbers.

Let

$$M^{00\ldots0} = \bigotimes_{i=1}^{n} (m_i^{-1} J_i), \tag{3.1.3}$$

which, together with (2.2.1), (2.2.2), defines M^x for every $x \in \Omega^*$. Also, let the $(v-1) \times v$ matrix P be defined as

$$P = (\ldots, P^{y'}, \ldots)', \tag{3.1.4}$$

where P^y is included in P for every $y \in \Omega$. For example, if $n = 2$ then $P = (P^{01'}, P^{10'}, P^{11'})'$.

Lemma 3.1.1. (a) For each $x \in \Omega^*$, Z^x can be expressed as a linear combination of M^y, $y \in \Omega^*$. (b) Conversely, for each $x \in \Omega^*$, M^x can be expressed as a linear combination of Z^y, $y \in \Omega^*$.

Proof. Follows immediately from (2.2.1), (2.2.2), (3.1.1), (3.1.2) and (3.1.3).

Lemma 3.1.2.

$$P' P = I - v^{-1} J,$$

where I is an identity matrix and J is a matrix of 1's both of order $v \times v$.

Proof. It may be seen from (2.2.5), (2.2.6), (2.2.8) ad (3.1.4) that in the v-dimensional Euclidian space, the rows of P form an orthonormal basis of the orthocomplement of the space of vectors having all elements equal. Hence the lemma follows. An alternative more explicit proof follows if one observes that by (2.2.7), (3.1.4),

$$P' P = \sum_{y \in \Omega} M^y,$$

and then applies induction on n.

Lemma 3.1.3. For a connected factorial design, (a) PCP' is positive definite (p.d.), (b) $P^y C P^{y'}$ is p.d. for every $y \in \Omega$.

Proof. Since C is non-negative definite (n.n.d.), for every (v-1)- component vector $\underline{u}$,

$$\underline{u}' PCP' \underline{u} \geq 0.$$

Furthermore, as the design is connected, equality holds in the above only if the elements of $P' \underline{u}$ are all equal, i.e., only if

$$P' \underline{u} = u_0 (\bigotimes_{i=1}^{n} \underline{1}_i), \tag{3.1.5}$$

for some constant u_0. By (2.2.6), (2.2.8), (3.1.4), PP' equals an identity matrix and $P(\bigotimes_{i=1}^{n} \underline{1}_i) = \underline{0}$. Hence, on pre-multiplication by P, (3.1.5) yields $\underline{u} = \underline{0}$. This proves (a). The proof of (b) now follows noting that for each $y \in \Omega$, $P^y CP^{y'}$ is a principal submatrix of PCP'.

Lemma 3.1.4. For a connected factorial design to be balanced with OFS, it is necessary and sufficient that the C-matrix of the design be of the form

$$C = \sum_{x \in \Omega} \rho(x) M^x, \tag{3.1.6}$$

where $\rho(x)$, $x \in \Omega$, are real numbers.

Proof. Sufficiency: Let C be of the form (3.1.6). Then by (2.2.7), (2.2.8), it is easy to see that for every $y, z \in \Omega$, $y \neq z$,

$$P^y C = \rho(y) P^y, \tag{3.1.7}$$

$$P^y CP^{y'} = \rho(y) I^{(y)}, \tag{3.1.8}$$

$$P^y CP^{z'} = 0, \tag{3.1.9}$$

where $I^{(y)}$ is the identity matrix of order $\Pi(m_i - 1)^{y_i}$ ($= \alpha(y)$, say). Note that the number of rows in P^y equals $\alpha(y)$.

By Lemma 3.1.3(b) and (3.1.8), $\rho(y) > 0$; hence by (3.1.7), $P^y = \{\rho(y)\}^{-1} P^y C$ and from the reduced normal equations $C\underline{\tau} = \underline{Q}$, it follows that the BLUE of $P^y \underline{\tau}$ is given by

$$P^y \hat{\underline{\tau}} = \{\rho(y)\}^{-1} P^y \underline{Q}, \quad y \in \Omega. \tag{3.1.10}$$

It is well-known that the dispersion matrix of $\underline{Q}$ is given by

$$Disp(\underline{Q}) = \sigma^2 C, \tag{3.1.11}$$

σ^2 being the constant error variance. Hence by (3.1.9), (3.1.10), for every $y, z \in \Omega$, $y \neq z$,

$$\begin{aligned} Cov(P^y\hat{\underline{\tau}}, P^z\hat{\underline{\tau}}) &= \{\rho(y)\rho(z)\}^{-1} Cov(P^y\underline{Q}, P^z\underline{Q}) \\ &= \sigma^2\{\rho(y)\rho(z)\}^{-1} P^y C P^{z'} = 0, \end{aligned}$$

which shows that the design has OFS. Also, by (3.1.8), (3.1.10), (3.1.11), for every $y \in \Omega$,

$$Disp(P^y\hat{\underline{\tau}}) = \sigma^2\{\rho(y)\}^{-1} I^{(y)}, \tag{3.1.12}$$

so that by Corollary 2.3.1, the design is balanced.

<u>Necessity:</u> Let P be as in (3.1.4). Since C has all row and column sums equal to zero, by Lemma 3.1.2,

$$P'PC = C = CP'P. \tag{3.1.13}$$

Hence $PC = PCP'P$. But by Lemma 3.1.3 (a), PCP' is p.d. Therefore, $P = (PCP')^{-1}PC$, and from the reduced normal equations $C\underline{\tau} = \underline{Q}$, the BLUE of $P\underline{\tau}$ is given by

$$P\hat{\underline{\tau}} = (PCP')^{-1} P\underline{Q}.$$

By (3.1.11),

$$Disp(P\hat{\underline{\tau}}) = \sigma^2 (PCP')^{-1}. \tag{3.1.14}$$

Suppose now that the design is balanced and has OFS. Since the design has OFS, $Cov(P^x\hat{\underline{\tau}}, P^y\hat{\underline{\tau}}) = 0$ for every $x, y\,(x \neq y) \in \Omega$. Hence all off-diagonal blocks in $Disp(P\hat{\underline{\tau}})$ must vanish so that by (3.1.14),

$$(PCP')^{-1} = \underset{x \in \Omega}{Diag}(\ldots, A_x, \ldots), \tag{3.1.15}$$

where $Disp(P^x\hat{\underline{\tau}}) = \sigma^2 A_x$, $x \in \Omega$. Since the design is balanced, by Corollary 2.3.1

for every $x \in \Omega$, A_x must be proportional to the identity matrix. Let $A_x = a_x I^{(x)}$, where $a_x > 0$. This, together with (3.1.15), yields

$$PCP' = \underset{x \in \Omega}{Diag}(\ldots, a_x^{-1} I^{(x)}, \ldots).$$

Pre- and post-multiplying the above by P' and P respectively, one obtains

$$C = \sum_{x \in \Omega} a_x^{-1} P^{x\prime} P^x = \sum_{x \in \Omega} a_x^{-1} M^x,$$

by (2.2.7), (3.1.4) and (3.1.13). Hence the necessity of the stated condition follows.

Theorem 3.1.1. For a connected factorial design to be balanced with OFS, it is necessary and sufficient that the C-matrix of the design has property A.

Proof. Necessity: This follows from Lemma 3.1.1. (b) and the necessity part of Lemma 3.1.4.

Sufficiency: Let the C-matrix have property A. Then by Lemma 3.1.1. (a), it is possible to express the C-matrix as

$$C = \sum_{x \in \Omega^*} \rho(x) M^x = \sum_{x \in \Omega} \rho(x) M^x + \rho(0,0,\ldots,0) M^{00\ldots0}, \tag{3.1.16}$$

where $\rho(x)$, $x \in \Omega^*$, are constants. By (2.2.1), (2.2.2), (3.1.3), for every $y \in \Omega$,

$$M^y M^{00\ldots0} = 0.$$

Also $M^{00\ldots0} M^{00\ldots0} = M^{00\ldots0}$ $(\neq 0)$ and $CM^{00\ldots0} = 0$, as each row sum of C equals zero. Hence, post-multiplying (3.1.16) by $M^{00\ldots0}$ it follows that $\rho(0,0,\ldots,0) = 0$. the sufficiency of the stated condition now follows from (3.1.16) and the sufficiency part of Lemma 3.1.4.

The above theorem provides a characterization for balance together with OFS in terms of property A of the C-matrix in the connected case. Hereafter, a design will be said to have property A if its C-matrix has property A. As indicated below, one can work out very simple formulae for the analysis of such designs.

Consider a connected design with property A. Then by Lemma 3.1.4 and Theorem 3.1.1, the BLUE of $P^y \tau$ and the dispersion matrix of this BLUE are given by (3.1.10) and (3.1.12) respectively. Hence by (2.2.7),

$$\begin{aligned} SS \quad \text{due to interaction} \quad F^{y} &= SS \quad \text{due to} \quad P^{y}\hat{\underline{\tau}} \\ &= (P^{y}\hat{\underline{\tau}})' \; [Disp\,(P^{y}\hat{\underline{\tau}})/\sigma^2]^{-1}(P^{y}\hat{\underline{\tau}}) \\ &= \{\rho(y)\}^{-1}\underline{Q}'P^{y\prime}P^{y}\underline{Q} \\ &= \{\rho(y)\}^{-1}\underline{Q}'M^{y}\underline{Q}, \quad y \in \Omega. \end{aligned} \tag{3.1.17}$$

The formula (3.1.17) is extremely simple in the sense that no matrix inversion is required.

The special cases of designs which are equireplicate or proper deserve some attention. For an equireplicate design with common replication number r,

$$C = r(\mathop{\otimes}_{i=1}^{n} I_i) - Nk^{-\delta}N',$$

and by Definition 3.1.1, C has property A if and only if $Nk^{-\delta}N'$ has property A. Similarly, if in addition, the design be proper with a constant block size k, then

$$C = r(\mathop{\otimes}_{i=1}^{n} I_i) - k^{-1}NN',$$

and C has property A if and only if NN' has property A. Hence our next result follows as a consequence of Theorem 3.1.1.

Theorem 3.1.2. (a) For a connected, equireplicate factorial design to be balanced with OFS, it is necessary and sufficient that the matrix $Nk^{-\delta}N'$ has property A. (b) For a connected, equireplicate, proper factorial design to be balanced with OFS, it is necessary and sufficient that the matrix NN' has property A.

For connected equireplicate designs with property A and a common replication number r, it is possible to give a simple formula for the interaction efficiencies. Recall that for the kind of design under consideration, the dispersion matrix of $P^{y}\hat{\underline{\tau}}$ is given by (3.1.12). On the other hand, it is readily seen that for a randomized (complete) block design with the same number of replicates, one would have obtained

$$Disp\,(P^{y}\hat{\underline{\tau}}) = \sigma^2 r^{-1} I^{(y)}. \tag{3.1.18}$$

A comparison between (3.1.12) and (3.1.18) shows that the efficiency with respect to interaction F^{y} in the design under consideration is given by, say,

$$\epsilon(y) = \rho(y)/r, \qquad y \in \Omega. \tag{3.1.19}$$

The following example serves as an illustration.

Example 3.1.1. Consider a 3×4 factorial arranged in twelve blocks as shown below.

<-- *BLOCKS* -->

00	00	00	01	01	01	02	02	02	03	03	03
11	12	13	10	12	13	10	11	13	10	11	12
22	23	21	23	20	22	21	23	20	22	20	21

The design is connected, proper with constant block size 3, and equireplicate with common replication number $r = 3$. It may be seen, by explicit computation, that for the above design,

$$\begin{aligned} NN' &= 3\, I_1 \otimes I_2 + (J_1 - I_1) \otimes (J_2 - I_2) \\ &= 4\, I_1 \otimes I_2 - I_1 \otimes J_2 - J_1 \otimes I_2 + J_1 \otimes J_2, \end{aligned}$$

where, as usual, I_1, I_2 are 3×3 and 4×4 identity matrices and J_1, J_2 are 3×3 and 4×4 matrices of all 1's. By (3.1.1), (3.1.2),

$$NN' = 4Z^{11} - Z^{10} - Z^{01} + Z^{11},$$

which shows that NN' has property A. Hence by Theorem 3.1.2 (b), the design is balanced and has OFS. Furthermore,

$$C = 3(I_1 \otimes I_2) - \frac{1}{3}NN' = \frac{5}{3}Z^{11} + \frac{1}{3}Z^{10} + \frac{1}{3}Z^{01} - \frac{1}{3}Z^{00},$$

Now, by (2.2.5), (2.2.6), (3.1.1), (3.1.2),

$$P^{10}Z^{11}P^{10'} = I^{(10)}, \quad P^{10}Z^{10}P^{10'} = 4I^{(10)},$$

$$P^{10}Z^{01}P^{10'} = P^{10}Z^{00}P^{10'} = 0.$$

Hence $P^{10}CP^{10'} = 3I^{(10)}$. Similarly, it may be seen that $P^{01}CP^{01'} = \frac{8}{3}I^{(01)}$, $P^{11}CP^{11'} = \frac{5}{3}I^{(11)}$. A comparison with (3.1.8) shows that $\rho(1,0) = 3$, $\rho(0,1) = \frac{8}{3}$,

$\rho(1,1) = \frac{5}{3}$. Hence by (3.1.19), the interaction efficiencies in the design under consideration are given by

$$\epsilon(1,0) = 1,\ \epsilon(0,1) = \frac{8}{9},\ \epsilon(1,1) = \frac{5}{9}.$$

As discussed in the next sections, the characterization for balance with OFS, as given in Theorem 3.1.1, has immediate applicability in the actual construction of designs. It is possible to derive another characterization, mainly of theoretical interest, in terms of the eigenvalues and eigenvectors of the C-matrix.

Theorem 3.1.3. For a connected factorial design to be balanced with OFS, it is necessary and sufficient that for every $x \in \Omega$, the columns of $P^{x'}$ represent an orthonormal system of eigenvectors corresponding to the same eigenvalue of C.

Proof. This is an immediate consequence of (2.2.7), (2.2.8) and Lemma 3.1.4.

We conclude this section with the remark that the major results considered here, namely Lemma 3.1.4 and Theorems 3.1.1, 3.1.3 are valid also for row-column designs and designs for multiway elimination of heterogeneity provided C is interpreted as the co-efficient matrix of the reduced normal equations for treatment effects in such designs (cf. Zelen and Federer (1964)).

3.2. A combinatorial characterization

Nair and Rao (1948) and Shah (1958, 1960a) defined a $(2^n - 1)$-class association scheme for n-factor experiments. Their association scheme has been referred to as extended group divisible (EGD) scheme by Hinkelmann and Kempthorne (1963) and binary number association scheme by Paik and Federer (1973). In this association scheme, two distinct treatment combinations are defined as x-th associates, $x \in \Omega$, where $x_i = 0$ if the ith factor occurs at the same level in both the treatment combinations and $x_i = 1$ otherwise. It is easy to see that the number of x-th associates of any treatment is then given by $\alpha(x) = \Pi(m_i - 1)^{x_i}$.

For $x \in \Omega$, let B^x be a $v \times v$ matrix such that its (j, j') th element equals 1 if the jth and j'th treatment combinations are x-th associates and zero otherwise. Then B^x, $x \in \Omega$, define the 2^n-1 association matrices of the EGD association scheme. Using the method of induction, it can be verified that (Gupta, 1988)

$$B^x = \bigotimes_{i=1}^{n} B_i^{x_i} \tag{3.2.1}$$

where for $1 \le i \le n$,

$$\left.\begin{aligned} B_i^{x_i} &= J_i - I_i \quad \text{if } x_i = 1 \\ &= I_i \qquad\quad\; \text{if } x_i = 0 \end{aligned}\right\} \tag{3.2.2}$$

<u>Definition 3.2.1.</u> An arrangement of the $v = \Pi m_i$ treatment combinations in b blocks each of size k will be called an extended group divisible (EGD) design if (i) the design is binary in the sense that each treatment combination occurs at most once in each block, (ii) each treatment combination occurs in exactly r blocks and (iii) every two distinct treatment combinations, which are x-th associates of each other, occur together in $\lambda(x)$ blocks, $x \in \Omega$.

For example, it may be seen that the design in Example 3.1.1. is an EGD design with parameters $n = 2$, $m_1 = 3$, $m_2 = 4$, $b = 12$, $r = k = 3$, $\lambda_{01} = \lambda_{10} = 0$, $\lambda_{11} = 1$.

It is readily seen (e.g. Raghavarao, 1971) that for an EGD design,

$$NN' = rI + \sum_{x \in \Omega} \lambda(x) B^x, \tag{3.2.3}$$

where I is the $v \times v$ identity matrix. By (3.1.1), (3.1.2), (3.2.2), for each $x \in \Omega$, B^x can be expressed as a linear combination of Z^y, $y \in \Omega^*$. Also, $I = Z^{11\ldots1}$. Hence by (3.2.3), for an EGD design, the matrix NN' has property A. Conversely, for each $x \in \Omega^*$, Z^x can be expressed as a linear combination of I and B^y, $y \in \Omega$. Consequently, if the NN' matrix of a binary proper design has property A, then the design must be an EGD design. We thus have the following result which was proved by Paik and Federer (1973) using an alternative argument.

Theorem 3.2.1. A binary, proper design is an EGD design if and only if its NN' matrix has property A.

Combining Theorems 3.1.2 (b), 3.2.1, one obtains the following result.

Theorem 3.2.2. For a connected, equireplicate, proper, binary factorial design to be balanced with OFS, it is necessary and sufficient that the design is an EGD design.

Theorem 3.2.2 presents a combinatorial characterization for balance with OFS in the case of connected, equireplicate, proper, binary designs. As will be seen later, such a characterization is of much help in the actual construction of designs. The 'sufficiency' part of this theorem was proved by Nair and Rao (1948), while the proof of the 'necessity' part is due to Shah (1958, 1960a) and Kshirsagar (1966). Incidentally, it appears that the indication of proof given here is simpler.

It is of interest to find explicit formulae for interaction efficiencies in an EGD design. To that effect, note that by (3.2.3), for an EGD design

$$C = k^{-1}\{r(k-1)I - \sum_{x \in \Omega} \lambda(x) B^x\}. \tag{3.2.4}$$

From (2.2.5), (2.2.6), (3.2.1), (3.2.2), it may be seen after a little algebra that for every $x, y \in \Omega$,

$$P^y B^x P^{y'} = [\prod_{i=1}^{n} \{(1-y_i)m_i - 1\}^{x_i}] I^{(y)}.$$

Hence by (3.2.4), $P^y C P^{y'} = \rho(y) I^{(y)}$, where

$$\rho(y) = k^{-1}[r(k-1) - \sum_{x \in \Omega} \lambda(x) \{\prod_{i=1}^{n} ((1-y_i)m_i - 1)^{x_i}\}]. \tag{3.2.5}$$

From (3.1.19), (3.2.5), the following result is evident.

Theorem 3.2.3. For a connected EGD design, with parameters as stated above, the efficiency with respect to the interaction F^y is given by

$$\epsilon(y) = k^{-1}[(k-1) - r^{-1} \sum_{x \in \Omega} \lambda(x) \{ \prod_{i=1}^{n} ((1-y_i)m_i - 1)^{x_i} \}].$$

3.3. A review of construction procedures

A brief account of the developments in constructions of balanced designs with OFS is presented in this section. Much of the work on this topic was done before 1970's, an account of which can be found also in Raghavarao (1971).

For symmetrical factorial experiments where the number of levels is prime or power of a prime, balanced designs with OFS can be obtained using the methods of Bose and Kishen (1940) and Bose (1947) by a suitable choice of a confounding scheme. Also, all balanced incomplete block designs for non-prime numbers of treatments are trivially balanced with OFS for some factorial experiment.

Using orthogonal arrays of strength 2, Nair and Rao (1948) gave methods for constructing EGD designs for an $m_1 \times m_2$ experiment in blocks of size m_1 or m_2. These authors also indicated how, starting from these, one can construct EGD designs involving more than 2 factors in simple cases. Following Nair and Rao (1948), several authors considered various methods for constructing these designs. Thomson and Dick (1951), starting from a basic $m_1 \times m_2$ design in blocks of size m_2 $(m_2 < m_1,$ m_1 being a prime number or power of a prime), obtained three factor designs with the same block size, and number of levels being m_1, m_2 or factors of m_2. Rao (1956) constructed some series of designs from orthogonal latin squares for $m_1 \times m_2$ experiments in blocks of size m_1 and $m_2 - 1$ replications. He also gave some designs for $2 \times m_2^2$ experiments. Kishen (1958) has given balanced designs with OFS of the type $m_1 \times 2^2$ and $m_1 \times m_2^2$. Kramer and Bradley (1957) and Zelen (1958) used group divisible incomplete block designs which have EGD scheme for two-factors. Kishen and Srivastava (1959) gave some general methods for constructing balanced designs with OFS for asymmetrical factorial experiments. They extended the method of finite geometries of Bose and Kishen (1940) by using curvilinear spaces or hypersurfaces and truncating the $EG(m,s)$ suitably, and by using vectors in Galois fields. They illustrated their methods by constructing the following series of designs: (i) $m_1^2 \times m_2$ design in blocks of size $m_1 m_2$, balanced in (m_1-1) replications $(m_1 > m_2)$, (ii) $m_1 \times m_2 \times m_3$ design in blocks of size

m_2m_3, balanced in (m_1-1) replications $(m_1 \geq m_2m_3)$, (iii) designs for experiments where the number of levels is a prime number or power of a prime number, (iv) $m_1 \times m_2 \times \cdots \times m_n$ design in blocks of size $m_1m_3m_4 \cdots m_n$ where m_2 is a factor of $m_1m_3m_4 \cdots m_n$ and is a prime number or power of a prime number $(m_2^2 \geq m_i, i \neq 2)$. Several other series of designs were also given by them. Das (1960) gave a method of construction for asymmetrical factorials by linking them with the fractions of suitable symmetrical factorials. Das and Giri (1986) have discussed this method in details and also gave several examples. Tharthare (1965) gave a class of balanced designs with OFS for $m_1 \times m_2^n$ experiments. Similar designs were obtained by Kishen and Tyagi (1964) using pairwise balanced designs of Bose and Shrikhande (1960). Shah (1960b) gave a method of construction for $m_1 + m_2$ factor experiment using balanced designs with OFS in m_1 and $m_2 + 1$ factors respectively. Shah (1960b) and Kishen and Tyagi (1963) constructed a 5×2^2 design in 10 blocks of size 2 each. An alternative design for this experiment can also be obtained using the method of Tharthare (1965). Muller (1966) developed designs for $m_1 \times m_2 \times \cdots \times m_n$ experiments where m_1 is prime or prime-power. His procedure is to replace each factor except the first by one or more pseudofactors each at m_1 levels. He also considered the use of balanced incomplete block designs for the construction of $m_1 \times m_2$ balanced factorials with OFS, when $m_1 > m_2$. Further construction procedures were suggested by Tyagi (1971) and Aggarwal (1974).

Among the more recent authors, Lewis and Tuck (1985) gave some designs with block size 2 while Gupta (1987c) presented an algorithm for obtaining a class of EGD designs (see also Chapter 5). Suen and Chakravarty (1986) constructed several series of two-factor balanced designs with OFS using balanced arrays of strength 2.

3.4. Concluding remarks.

The results presented in this chapter relate to connected factorial designs. The disconnected case poses some special problems. In particular, then the conclusions of Lemma 3.1.3, which is helpful in proving the subsequent results, no more remain valid. Of course, one may work with generalized inverses of matrices but even then some special considerations are required. As a matter of fact, the results proved in this chapter, at least in their present forms, do not remain valid in the disconnected case. The following example illustrates the point.

Example 3.4.1. Consider a disconnected 2^3 design in two blocks as shown below.

Block I: 000, 100, 010, 001

Block II: 110, 101, 011, 111

Clearly, each interaction is represented by a single contrast. It may be seen from elementary considerations that the contrasts belonging to interactions F^{110}, F^{101}, F^{011} are estimable while those belonging to F^{100}, F^{010}, F^{001}, F^{111} are not estimable. Moreover, BLUE's of the contrasts belonging to F^{110}, F^{101}, F^{011} may be seen to be mutually orthogonal, i.e., uncorrelated. Hence the design has OFS. Also, trivially the design is balanced since each interaction is represented by a single contrast. Thus the design is balanced and has OFS. However, the C-matrix is not of the form (3.1.6). In order to appreciate this point, note that if the C-matrix be of the form (3.1.6), then by (2.2.7), (2.2.8), one must have

$$M^y C = CM^y$$

for every $y \in \Omega$. For this design, explicit computation shows that, in particular, M^{001} does not commute with C.

The above example demonstrates that in the disconnected case the necessity part of Lemma 3.1.4 does not necessarily remain valid. Similarly, it may be shown that the necessity parts of Theorems 3.1.1, 3.1.2, 3.1.3, 3.2.2 may not remain valid for disconnected designs. This calls for suitable modifications of these results to make them applicable to the disconnected case. The relevant details will be taken up in the next chapter.

CHAPTER 4
CHARACTERIZATIONS FOR ORTHOGONAL FACTORIAL STRUCTURE

4.1. Introduction and preliminaries

In Chapter 3, we have considered several characterizations for factorial designs which are balanced and have OFS. The emphasis on balance, however, has a drawback that the resulting designs, although theoretically elegant and simple to interpret, may become too large and hence expensive. More specifically, the combinatorial restrictions imposed by an attempt to achieve a balance over all the interactions (cf Theorem 3.2.2) may call for a prohibitively large number of replications. Because of this reason, since the early seventies, work started on conditions for OFS alone. A brief review of these developments up to that stage was given by Chatterjee (1982). In this chapter, we propose to present a comprehensive account of several characterizations for OFS considering both connected and disconnected designs. The use of these results for construction purposes will be indicated later in the monograph. Before presenting the main results of this chapter, we introduce some preliminaries that will be useful in the subsequent development.

A permutation matrix is a square matrix with exactly one entry in each row and column equal to 1 and the rest equal to zero. A proper matrix is a square matrix with all row and column sums equal.

Definition 4.1.1. A $v \times v$ matrix G, where $v = \Pi m_i$, will be said to have structure K if it can be expressed as a linear combination of Kronecker products of permutation matrices of orders $m_1, m_2, \ldots, m_n$, i.e., if

$$G = \sum_{g=1}^{w} \xi_g \left(\bigotimes_{i=1}^{n} R_{gi} \right),$$

where w is some positive integer, $\xi_1, \xi_2, \ldots, \xi_w$ are some real numbers, and for each g, R_{gi} is some permutation matrix of order m_i $(1 \leq i \leq n)$.

Note that every permutation matrix is a proper matrix. Conversely, it can be shown (see e.g., Roberts and Verlag (1973, pp. 200)) that every proper matrix can be expressed as a linear combination of permutation matrices of the same order. Hence one obtains the following equivalent version of Definition 4.1.1.

Definition 4.1.1A. A $v \times v$ matrix G, where $v = \Pi m_i$, will be said to have structure K if it can be expressed as a linear combination of Kronecker products of proper matrices of orders $m_1, m_2, \ldots, m_n$.

4.2. Algebraic characterizations: the connected case

First we prove two important lemmas.

Lemma 4.2.1. For every $v \times v$ permutation matrix R and every $x \in \Omega$, $M^x R M^x$ has structure K.

Proof. We label the columns of M^x by an n-tuplet of subscripts $(j_1, j_2, \ldots, j_n)$, $0 \le j_i \le m_i - 1$, $1 \le i \le n$, or by a single subscript j, $1 \le j \le v$, according to convenience. Denoting the $(j_1, j_2, \ldots, j_n)$th column of M^x by $\underline{h}_{j_1 j_2 \cdots j_n}$, by (2.2.1), (2.2.2), we have

$$\underline{h}_{j_1 j_2 \cdots j_n} = \bigotimes_{i=1}^{n} \underline{h}_{ij_i}, \tag{4.2.1}$$

where $\underline{h}_{ij_i}$ is the j_ith column of $M_i^{x_i}$, $1 \le i \le n$. Now labelling the columns by a single subscript, let M^x be of the form $M^x = (\underline{\gamma}_{11}, \ldots, \underline{\gamma}_{1v})$. Since R is a permutation matrix, $M^x R$ is of the form $M^x R = (\underline{\gamma}_{21}, \ldots, \underline{\gamma}_{2v})$, the columns of $M^x R$ being obtained by permuting the columns of M^x. Since $M^x R M^x = \sum_{j=1}^{v} \underline{\gamma}_{2j} \underline{\gamma}_{1j}'$, it is enough to show that for every j, $\underline{\gamma}_{2j} \underline{\gamma}_{1j}'$ has structure K. Now,

$$\underline{\gamma}_{2j} \underline{\gamma}_{1j}' = \underline{h}_{j_1 j_2 \cdots j_n} \underline{h}_{j_1' j_2' \cdots j_n'}',$$

for some $(j_1, j_2, \ldots, j_n)$ and $(j_1', j_2', \ldots, j_n')$, not necessarily identical. Hence by (4.2.1),

$$\underline{\gamma}_{2j} \underline{\gamma}_{1j}' = \bigotimes_{i=1}^{n} (\underline{h}_{ij_i} \underline{h}_{ij_i'}'). \tag{4.2.2}$$

If $x_i = 0$, then $M_i^{x_i} = m_i^{-1} J_i$, so that $\underline{h}_{ij_i} = \underline{h}_{ij_i'} = m_i^{-1} \underline{1}_i$ and

$$\underline{h}_{ij_i}\underline{h}'_{ij_i'} = m_i^{-2}J_i,$$

which is a proper matrix. Similarly, if $x_i = 1$, then by (2.2.2),

$$\underline{h}_{ij_i} = \underline{e}_{ij_i} - m_i^{-1}\underline{1}_i, \ \underline{h}_{ij_i'} = \underline{e}_{ij_i'} - m_i^{-1}\underline{1}_i,$$

where $\underline{e}_{il}$ is an $m_i \times 1$ vector having 1 at the lth position $(0 \le l \le m_i - 1)$ and 0 elsewhere, and it may be readily seen that $\underline{h}_{ij_i}\underline{h}'_{ij_i'}$ is again a proper matrix. Hence by (4.2.2), for each j, $\underline{\gamma}_{2j}\underline{\gamma}'_{1j}$ can be expressed as a Kronecker product of proper matrices of orders $m_1, m_2, \ldots, m_n$ and, as such, has structure K.

Lemma 4.2.2. For each $x \in \Omega$, $M^x CM^x$ has structure K, where C is the usual C-matrix of a design.

Proof. The matrix C has all row and column sums equal to zero and is, therefore, proper. Hence C can be expressed as a linear combination of permutation matrices. The lemma now follows from the preceding one.

We now present the main result of this section.

Theorem 4.2.1 (Mukerjee (1979)). For a connected factorial design to have OFS, it is necessary and sufficient that the C-matrix of the design has structure K.

Proof. Sufficiency: From (2.2.2), observe that for $1 \le i \le n$, if R_i be an $m_i \times m_i$ permutation matrix then $M_i^{x_i}R_i = R_iM_i^{x_i}$ $(x_i = 0,1)$. Hence by (2.2.1) and Definition 4.1.1, if the C-matrix has structure K, then for every $x \in \Omega$,

$$M^x C = CM^x. \tag{4.2.3}$$

Consequently, by (2.2.7), (2.2.8), $P^x C = P^x CP^{x\prime} P^x$, and by Lemma 3.1.3(b), it follows that $P^x = (P^x CP^{x\prime})^{-1}P^x C$. Hence from the reduced normal equations $C\underline{\tau} = \underline{Q}$, one obtains the BLUE of $P^x\underline{\tau}$ as

$$P^x\hat{\underline{\tau}} = (P^x CP^{x\prime})^{-1}P^x\underline{Q}, \quad x \in \Omega. \tag{4.2.4}$$

Since $Disp(\underline{Q}) = \sigma^2 C$, by (4.2.4) for every x, $y (x \ne y) \in \Omega$,

$$Cov(P^x\hat{\tau}, P^y\hat{\tau}) = \sigma^2(P^x CP^{x'})^{-1}P^x CP^{y'}(P^y CP^{y'})^{-1}. \tag{4.2.5}$$

But by (2.2.7), (2.2.8), (4.2.3),

$$\begin{aligned} P^x CP^{y'} &= P^x P^{x'} P^x CP^{y'} = P^x M^x CP^{y'} \\ &= P^x CM^x P^{y'} = P^x CP^{x'} P^x P^{y'} = 0, \end{aligned}$$

whenever $x \neq y$. Therefore, by (4.2.5), $Cov(P^x\hat{\tau}, P^y\hat{\tau}) = 0$ for every x, $y(x \neq y) \in \Omega$, which shows that the design has OFS.

Necessity: If the design has OFS, then defining the $(v-1) \times v$ matrix P by (3.1.4) and proceeding as in the proof of the necessity part of Lemma 3.1.4,

$$(PCP')^{-1} = \underset{x \in \Omega}{Diag}(\ldots, A_x, \ldots),$$

where $Disp(P^x\hat{\tau}) = \sigma^2 A_x$, $x \in \Omega$. Hence

$$PCP' = Diag(\ldots, A_x^{-1}, \ldots). \tag{4.2.6}$$

Note that the existence of A_x^{-1}, $x \in \Omega$, is guaranteed as PCP' is p.d. by Lemma 3.1.3 (a). Pre- and post-multiplying (4.2.6) by P' and P respectively, one obtains

$$C = \sum_{x \in \Omega} P^{x'} A_x^{-1} P^x, \tag{4.2.7}$$

by (3.1.4) and (3.1.13). By (2.2.8), (4.2.7), for every $x \in \Omega$, $P^x CP^{x'} = A_x^{-1}$, and this, together with (2.2.7), (4.2.7) implies that

$$C = \sum_{x \in \Omega} P^{x'} P^x CP^{x'} P^x = \sum_{x \in \Omega} M^x CM^x.$$

From Lemma 4.2.2, it is now immediate that C must have structure K.

Considering the special cases of designs which are equireplicate or proper, one has the following result analogously to Theorem 3.1.2.

Theorem 4.2.2. (a) For a connected, equireplicate factorial design to have OFS, it is necessary and sufficient that the matrix $Nk^{-\delta}N'$ as structure K. (b) For a connected, equireplicate, proper factorial design to have OFS, it is necessary and sufficient that the matrix NN' has structure K.

The necessary and sufficient condition obtained in Theorem 4.2.1 can be expressed in alternative forms, as shown in the following theorems, which may be helpful in checking whether a given connected factorial design has OFS or not.

Theorem 4.2.3. For a connected factorial design to have OFS, it is necessary and sufficient that the C-matrix satisfies

$$C = \sum_{x \in \Omega} M^x C M^x. \tag{4.2.8}$$

Proof. The 'necessity' part has already been proved while proving Theorem 4.2.1. The sufficiency of (4.2.8) follows from Lemma 4.2.2 and Theorem 4.2.1.

Theorem 4.2.4. For a connected factorial design to have OFS, it is necessary and sufficient that for every $x \in \Omega$, M^x commutes with C.

Proof. The 'necessity' part is a consequence of (2.2.7), (2.2.8) and Theorem 4.2.3. The proof of the 'sufficiency' part is contained in the proof of Theorem 4.2.1.

Theorem 4.2.5. For a connected factorial design to have OFS, it is necessary and sufficient that for every $x, y \in \Omega$, $x \neq y$, $M^x C M^y = 0$.

Proof. Sufficiency: By (2.2.7), (3.1.4), and Lemma 3.1.2,

$$\sum_{x \in \Omega} M^x = P'P = I - v^{-1}J.$$

Since all row and column sums of C equal zero, one obtains

$$\begin{aligned} C &= (I - v^{-1}J)C(I - v^{-1}J) = (\sum_{x \in \Omega} M^x)C(\sum_{x \in \Omega} M^x) \\ &= \sum_{x \in \Omega} \sum_{y \in \Omega} M^x C M^y = \sum_{x \in \Omega} M^x C M^x, \end{aligned}$$

under the stated condition, and from the 'sufficiency' part of Theorem 4.2.3, it follows that the design has OFS.

Necessity: Follows from (2.2.7), (2.2.8), and Theorem 4.2.4.

Theorem 4.2.6. For a connected factorial design to have OFS, it is necessary and sufficient that for every $x, y \in \Omega$, $x \neq y$, $P^x C P^{y'} = 0$.

Proof. This is an immediate consequence of (2.2.7), (2.2.8) and Theorem 4.2.5.

Theorem 4.2.7. For a connected factorial design to have OFS, it is necessary and sufficient that for every $x \in \Omega$, Z^x commutes with C.

Proof. Follows from Lemma 3.1.1, Theorem 4.2.4 and the fact that all row and column sums of C equal zero.

Let $\mu(A)$ denote the column space of any matrix A. Then, analogously to Theorem 3.1.3, the following result holds (cf. Bailey (1985a)).

Theorem 4.2.8. For a connected factorial design to have OFS, it is necessary and sufficient that for every $x \in \Omega$, the C-matrix has $\alpha(x) = \Pi(m_i - 1)^{x_i}$ orthonormal eigenvectors belonging to $\mu(P^{x'})$.

Proof. Necessity: If the design has OFS, then by Theorem 4.2.4, for every $x \in \Omega$, C and M^x commute and hence by well-known results (see e.g., Rao (1973a, pp. 41)) have v common orthonormal eigenvectors. Since by (2.2.1), (2.2.2), rank $(M^x) = \alpha(x)$ and M^x is n.n.d., of these eigenvectors $\alpha(x)$ correspond to positive eigenvalues of M^x and hence belong to $\mu(M^x)$. By (2.2.7), $\mu(M^x) \equiv \mu(P^{x'})$ and the 'necessity' part follows.

Sufficiency: For each $x \in \Omega$, let C have $\alpha(x)$ orthonormal eigenvectors, say $\underline{a}_1^x, \ldots, \underline{a}_{\alpha(x)}^x$, belonging to $\mu(P^{x'})$ and corresponding to the eigenvalues $\lambda_1^x, \ldots, \lambda_{\alpha(x)}^x$ (not necessarily distinct) of C. Let $\Lambda^x = Diag(\lambda_1^x, \ldots, \lambda_{\alpha(x)}^x)$, $L^x = (\underline{a}_1^x, \ldots, \underline{a}_{\alpha(x)}^x))$. Then

$$CL^x = L^x \Lambda^x. \tag{4.2.9}$$

Since rank $(P^{x'}) = \alpha(x)$, the columns of L^x form an orthonormal basis of $\mu(P^{x'})$ and from (2.2.7), it may be seen that $M^x = L^x L^{x'}$. Hence by (4.2.9), for each $x \in \Omega$,

$$M^* C = L^* L^{*\prime} C = L^* \Lambda^* L^{*\prime} = C L^* L^{*\prime} = C M^*,$$

and the design has OFS by Theorem 4.2.4.

All the theorems presented so far in this section are in terms of the C-matrix of the design. Some further results, in terms of a generalized (g-) inverse of the C-matrix, have also been reported in the literature. John and Smith (1972) obtained a useful sufficient condition for OFS in a two-factor setting. Their result was generalized by Cotter, John and Smith (1973) to the general n-factor situation. For $0 \leq j_i \leq m_i - 1$, $1 \leq i \leq n$, let $R^*_{ij_i}$ be an $m_i \times m_i$ circulant matrix with the first row having unity in the $j_i\,th$ position and zero elsewhere. For example, if $m_i = 3$, then

$$R^*_{i0} = \begin{pmatrix} 1 & 0 & 0 \\ 0 & 1 & 0 \\ 0 & 0 & 1 \end{pmatrix}, \; R^*_{i1} = \begin{pmatrix} 0 & 1 & 0 \\ 0 & 0 & 1 \\ 1 & 0 & 0 \end{pmatrix}, \; R^*_{i2} = \begin{pmatrix} 0 & 0 & 1 \\ 1 & 0 & 0 \\ 0 & 1 & 0 \end{pmatrix}.$$

The following result was proved by Cotter, John and Smith (1973).

Theorem 4.2.9. Let the C-matrix of a connected factorial design have a g-inverse of the form

$$\overline{C} = \sum_{j_1=0}^{m_1-1} \cdots \sum_{j_p=0}^{m_p-1} \left(\bigotimes_{i=1}^{p} R^*_{ij_i} \right) \otimes G_{j_1 j_2 \cdots j_p},$$

where $p = n - 1$ and for each $j_1, j_2, \ldots, j_p$, $G_{j_1 j_2 \cdots j_p}$ is a proper matrix of order m_n. Then the design has OFS.

Proof. Since the design is connected, for each $x \in \Omega$, $R(M^x) \subset R(C)$ where, as before, $R(A)$ represents the row space of any matrix A. Hence

$$M^x = M^x \overline{C} C,$$

and from the reduced normal equations $C\underline{\tau} = \underline{Q}$, the BLUE of $M^x \underline{\tau}$ is given by $M^x \underline{\hat{\tau}} = M^x \overline{C} \underline{Q}$. As $Disp\ (\underline{Q}) = \sigma^2 C$, it follows that for every $x, y \in \Omega$, $x \neq y$,

$$Cov(M^x\hat{\tau}, M^y\hat{\tau}) = \sigma^2 M^x \overline{C}\ C\ \overline{C}' M^y = \sigma^2 M^x \overline{C}' M^y. \tag{4.2.10}$$

Clearly, $\overline{C}'$ has structure K and hence, as in the proof of Theorem 4.2.1, $M^x \overline{C}' = \overline{C}' M^x$, for each $x \in \Omega$. Consequently, by (2.2.7), (2.2.8), (4.2.10), $Cov(M^x\hat{\tau}, M^y\hat{\tau}) = 0$, for every x, $y \in \Omega$, $x \neq y$.

Cotter, John and Smith (1973) proved the above result by induction on n. The present proof appears to be simpler.

Before concluding this section, we present a formula for the analysis of a connected factorial design with OFS. As shown earlier, the C-matrix of such a design has structure K and hence the BLUE of $P^x\tau$ is given by (4.2.4). Clearly,

$$Disp(P^x\hat{\tau}) = \sigma^2(P^x CP^{x\prime})^{-1},$$

and, analogously to (3.1.7), it may be seen that

$$SS \ \ due\ to\ interaction\ \ F^x = Q' P^{x\prime}(P^x CP^{x\prime})^{-1} P^x Q, \ \ x \in \Omega. \tag{4.2.11}$$

The formula (4.2.11), although not as simple as (3.1.7), is not too complex either. In particular, it does not require inversion of large matrices in the sense that although the C-matrix is of order $v \times v$, an application of (4.2.11) involves inversion of matrices only of order $\alpha(x)$, $x \in \Omega$.

4.3. Algebraic characterizations: the disconnected case

In Chapter 3, it was seen that the disconnected case poses some special problems in characterizing designs which are balanced with OFS. Similar problems exist even in characterizing designs having OFS alone. This is illustrated by Example 3.4.1, where the disconnected design under consideration has OFS but still then the C-matrix of the design does not commute with M^{001} and hence does not have structure K. It follows that the results in the preceding section, at least in their present forms, do not remain valid in the disconnected case.

In order to handle disconnected designs in an effective manner, we bring in the notion of regularity, defined as follows. For $x \in \Omega$, let

$$V_x \equiv R(P^x), \ \ V_x^* = R(P^x) \cap R(C). \tag{4.3.1a}$$

Note that any contrast belonging to F^x is estimable if and only if the corresponding coefficient vector belongs to $R(C)$. Since $P^x \underline{\tau}$ represents a complete set of orthonormal contrasts belonging to F^x, V_x^* is the space of coefficient vectors of estimable contrasts belonging to F^x. From (4.3.1a), it is easily seen that.

$$R(C) \supset \bigoplus_{x \in \Omega} V_x^*, \tag{4.3.1b}$$

where $\oplus$ denotes direct sum (see e.g., Rao (1973a, pp. 11)).

Definition 4.3.1. A disconnected factorial design will be called regular if $\bigoplus_{x \in \Omega} V_x^*$ equals $R(C)$ and irregular if it is a proper subspace of $R(C)$.

It may be remarked that in a connected design all treatment contrasts are estimable and hence $V_x^* \equiv V_x$, for each $x \in \Omega$. It is then readily seen that $R(C) \equiv \bigoplus_{x \in \Omega} V_x^*$, so that a connected design is always regular.

In a regular factorial design, under OFS the adjusted SS due to treatments (i.e., due to all estimable treatment contrasts) can be split up orthogonally into components corresponding to BLUE's of estimable contrasts belonging to the interactions only. In irregular designs, however, estimable contrasts belonging to interactions do not span the space of all estimable treatment contrasts and the adjusted treatment SS always contains a component due to estimable treatment contrasts belonging to none of the interactions.

Example 4.3.1. Consider the disconnected 2^3 design presented in Example 3.4.1. It may be shown that for this design

$$V_{100}^* \equiv V_{010}^* \equiv V_{001}^* \equiv V_{111}^* \equiv \{\underline{0}\},$$

$$V_{110}^* \equiv V_{110},\ V_{101}^* \equiv V_{101},\ V_{011}^* \equiv V_{011}.$$

Hence,

$$rank(\bigoplus_{x \in \Omega} V_x^*) = rank(V_{110} \oplus V_{101} \oplus V_{011}) = 3 < 6 = rank(C).$$

Thus this is an irregular disconnected design. It is easy to check that the estimable treatment contrasts belonging to interactions do not span the space of all estimable treatment contrasts. For example, none of the contrasts $P^{100}\underline{\tau}$, $P^{010}\underline{\tau}$, $P^{001}\underline{\tau}$, $P^{111}\underline{\tau}$ representing interactions F^{100}, F^{010}, F^{001}, F^{111} respectively, are estimable, although the contrast $(P^{111} + P^{001} + P^{010} - P^{100})\underline{\tau}$, obtained as their linear combination, is estimable. As already seen, the design under consideration has OFS. Hence, it follows that for disconnected designs OFS does not necessarily imply regularity.

Since for irregular designs an exhaustive orthogonal splitting of the adjusted treatment SS into components corresponding to different interactions is not possible, hereafter irregular disconnected designs are left out of consideration. In fact, in the factorial context, an irregular design which achieves the estimability of some unimportant treatment contrasts at the cost of the more important ones (i.e., those belonging to the interactions), is uneconomical to use.

Necessary and sufficient conditions for OFS in disconnected factorial designs are derived in the following theorems (Mukerjee, 1979).

Theorem 4.3.1. For a regular disconnected factorial design to have OFS, it is necessary that the C-matrix of the design has structure K.

Proof. For each $x \in \Omega$, let H_x be a matrix such that the rows of H_x form an orthonormal basis of V_x^*. Then

$$H_x = S_x P^x , \tag{4.3.2}$$

say. The spaces V_x^* are mutually orthogonal since so are the spaces V_x. Let H be a matrix such that (cf.(3.1.4)),

$$H = (\ldots, \underset{x \in \Omega}{H_x'}, \ldots)' .$$

Since the design is regular, the rows of H form an orthonormal basis of $\bigoplus_{x \in \Omega} V_x^*$ and hence of $R(C)$. By standard results $H'HCH'H = C$, $H'H$ commutes with C and HCH' is p.d. For each $x \in \Omega$, $H_x C H_x'$, being a principal submatrix of HCH', is p.d.

Hence, if the design has OFS, then noting that for every $x, y \in \Omega$, $x \neq y$, $Cov(H_x \hat{\underline{\tau}}, H_y \hat{\underline{\tau}}) = 0$, and proceeding exactly as in the 'necessity' part of Theorem 4.2.1, we have in analogy with (4.2.8),

$$C = \sum_{x \in \Omega} H_x' H_x C H_x' H_x. \tag{4.3.3}$$

Now, for $x \in \Omega$, by (2.2.7), (2.2.8), (4.3.2),

$$H_x' H_x = P^{x\prime} S_x' S_x P^x = M^x P^{x\prime} S_x' S_x P^x M^x = M^x H_x' H_x M^x. \tag{4.3.4}$$

From (4.3.2), it is easy to check that $H_x' H_x$ is a proper matrix with each row and column sum equal to zero, so that $H_x' H_x$ can be expressed as a linear combination of permutation matrices of the same order. Using such an expression for $H_x' H_x$ in the right-hand member of (4.3.4) and applying Lemma 4.2.1, it follows that $H_x' H_x$ has structure K. From (4.3.3), (4.3.4),

$$C = \sum_{x \in \Omega} M^x (H_x' H_x)(M^x C M^x)(H_x' H_x) M^x. \tag{4.3.5}$$

For each $x \in \Omega$, by definition M^x has structure K. Also by Lemma 4.2.2, $M^x C M^x$ has structure K. Since for each $x \in \Omega$, M^x, $H_x' H_x$, $M^x C M^x$ have structure K, the result follows from (4.3.5).

<u>Remark.</u> In the above proof it has been assumed implicitly that $V_x^* \not\equiv \{ \underline{0} \}$, for every $x \in \Omega$. If $V_x^* \equiv \{ \underline{0} \}$ for some x, the proof will remain similar, the only change being that the corresponding H_x matrix will not occur in H.

<u>Theorem 4.3.2.</u> If the C-matrix of a disconnected factorial design has structure K then the design (i) is regular, (ii) has OFS.

<u>Proof.</u> (i) If C has structure K, then as seen in the proof of the 'sufficiency' part of Theorem 4.2.1, for each $x \in \Omega$, C and M^x commute, i.e., (4.2.3) holds. hence by (2.2.7), (2.2.8),

$$P^x C = P^x M^x C = P^x C M^x = P^x C P^{x\prime} P^x,$$

so that

$$R(P^x C) \subset R(P^x) \equiv V_x. \tag{4.3.6}$$

Also, $R(P^x C) \subset R(C)$, which, together with (4.3.6), yields

$$R(P^x C) \subset R(P^x) \cap R(C) \equiv V_x^*. \tag{4.3.7}$$

Since C has all row and column sums equal to zero, by (3.1.4), (4.3.7) and Lemma 3.1.2,

$$R(C) \equiv R((I - v^{-1}J)C) \equiv R(P'PC) \subset R(PC) \equiv \underset{x \in \Omega}{\oplus} R(P^x C) \subset \underset{x \in \Omega}{\oplus} V_x^*, .$$

which, combined with (4.3.1b), gives $R(C) \equiv \underset{x \in \Omega}{\oplus} V_x^*$, i.e., the design is regular.

(ii) For each $x \in \Omega$, let H_x be as in the proof of the preceding theorem. Since $R(H_x) \subset R(C)$, there exists a matrix U_x such that $H_x = U_x C$. By (4.3.2), $U_x C = S_x P^x$, and hence by (2.2.8), $S_x = U_x C P^{x'}$. Therefore, by (2.2.7), (4.3.2),

$$H_x = S_x P^x = U_x C P^{x'} P^x = U_x C M^x. \tag{4.3.8}$$

As already noted, under structure K of C, for each $x \in \Omega$, C and M^x commute. Thus by (4.3.8), $H_x = U_x M^x C$, and from the reduced normal equations $C\underline{\tau} = \underline{Q}$, the BLUE of $H_x \underline{\tau}$ is given by

$$H_x \underline{\hat{\tau}} = U_x M^x \underline{Q}, \tag{4.3.9}$$

whence it may be seen that for every $x, y \in \Omega$, $x \neq y$, $Cov(H_x \underline{\hat{\tau}}, H_y \underline{\hat{\tau}}) = 0$, exactly as in the 'sufficiency' part of Theorem 4.2.1.

Theorems 4.3.1, 4.3.2 extend Theorem 4.2.1 to the disconnected case. In a similar manner, it may be seen that the conclusions of Theorems 4.2.3 - 4.2.8 remain valid even in the disconnected case provided the phrase "for a connected factorial design to have OFS" in the statements of these theorems be replaced by "for a disconnected factorial design to be regular and have OFS". A similar remark holds good for Theorem 4.2.2 as well.

<u>Example 4.3.2.</u> Consider a 2×4 factorial design in eight blocks as shown below.

<---			Blocks		--->		
00	01	02	03	01	02	03	00
11	12	13	10	10	11	12	13

The design is disconnected and for this design it may be checked that

$$NN' = 2\begin{pmatrix}1 & 0\\0 & 1\end{pmatrix} \otimes \begin{pmatrix}1&0&0&0\\0&1&0&0\\0&0&1&0\\0&0&0&1\end{pmatrix} + \begin{pmatrix}0 & 1\\1 & 0\end{pmatrix} \otimes \begin{pmatrix}0&1&0&1\\1&0&1&0\\0&1&0&1\\1&0&1&0\end{pmatrix}.$$

Thus NN' can be expressed as a linear combination of Kronecker products of proper matrices of orders 2 and 4, and hence has structure K. Since the design is equireplicate and proper, its C-matrix will also have structure K. As a result, by Theorem 4.3.2, the design is regular and has OFS. In fact, for the above design, the C-matrix has rank 6. All contrasts belonging to the main effects and two independent contrasts belonging to the two-factor interaction are estimable. This accounts for all the 6 degrees of freedom carried by estimable treatment contrasts.

It is of interest to extend the formula (4.2.11) for sum of squares to the disconnected case. The following lemma will be helpful in this context.

Lemma 4.3.1. If the C-matrix of a disconnected factorial design has structure K, then for every $x \in \Omega$, $R(P^x C) \equiv V_x^*$.

Proof. Take any vector $\underline{z}_1 \in V_x^* (\equiv R(P^x) \cap R(C))$. Clearly, there exist vectors $\underline{z}_2$, $\underline{z}_3$ such that

$$\underline{z}_1' = \underline{z}_2' C = \underline{z}_3' P^x. \tag{4.3.10}$$

Since under the stated condition, $M^x (=P^{x\prime} P^x$, by (2.2.7)) commutes with C, by (2.2.8), (4.3.10), one obtains

$$\underline{z}_1' = \underline{z}_3' P^x = \underline{z}_3' P^x P^{x\prime} P^x = \underline{z}_2' C P^{x\prime} P^x = \underline{z}_2' P^{x\prime} P^x C,$$

so that $\underline{z}_1' \in R(P^x C)$. Hence, $V_x^* \subset R(P^x C)$. Again, as in the proof of Theorem 4.3.2(i), under the stated condition, $R(P^x C) \subset V_x^*$. Hence the lemma follows.

The above lemma shows that if the C-matrix of a disconnected factorial design has structure K, then $P^x C\underline{\tau}$ represents a complete set of estimable treatment contrasts belonging to interaction F^x, $x \in \Omega$. From the reduced normal equations $C\underline{\tau} = \underline{Q}$, the BLUE of $P^x C\underline{\tau}$ is given by $P^x C\hat{\underline{\tau}} = P^x \underline{Q}$, with Disp

$(P^x C\underset{\sim}{\tau}) = \sigma^2 P^x C P^{x'}$. Hence if the C-matrix of a disconnected factorial design has structure K (or equivalently, by Theorems 4.3.1, 4.3.2, if the design be regular with OFS), then analogously to (4.2.11),

$$\begin{aligned} SS \text{ due to interaction } F^x &= SS \text{ due to } P^x C\underset{\sim}{\tau} \\ &= \underset{\sim}{Q}' P^{x'} (P^x C P^{x'})^- P^x \underset{\sim}{Q}, \quad x \in \Omega, \qquad (4.3.11) \end{aligned}$$

where $(P^x C P^{x'})^-$ is any g-inverse of $P^x C P^{x'}$. By Lemma 4.3.1, the degree(s) of freedom (d.f.) carried by the SS (4.3.11) equals rank $(P^x C)$.

While concluding this section, we indicate the modifications of the results in Chapter 3 required to make them valid for the disconnected case. Proceeding along the lines of proofs of Theorems 4.3.1, 4.3.2, one may obtain the following version of Theorem 3.1.1 in the context of disconnected designs.

Theorem 4.3.3. For a regular disconnected design to be balanced with OFS, it is necessary that the C-matrix of the design has property A. Conversely, if the C-matrix has property A, then the design is (i) regular and (ii) balanced with OFS.

Similar modifications of Theorems 3.1.2, 3.1.3, and 3.2.1 are easy to work out.

4.4. Partial orthogonal factorial structure

In factorial experimentation, especially with a large number of factors, it is sometimes the case that the higher order interactions are considered unimportant or of little practical interest. In such a situation, a design with complete OFS, in the sense of the preceding sections, is unnecessary. All that is required is a design which admits orthogonality with respect to the lower order interactions. Also, designs with partial orthogonal factorial structure will be recommended when appropriate designs with complete OFS are not available. This section considers the corresponding characterization problems. As before, connected designs are considered first and disconnected designs are taken up at a later stage.

Definition 4.4.1. For a fixed $x \in \Omega$, a factorial design will be said to have partial orthogonal factorial structure (POFS) with respect to the interaction F^x if the

BLUE's of estimable contrasts belonging to F^x are orthogonal to, i.e., uncorrelated with the BLUE's of estimable contrasts belonging to every other interaction.

The following theorem, due to Chauhan and Dean (1986), gives a characterization for POFS with respect to F^x in the connected case.

Theorem 4.4.1. For a connected factorial design to have POFS with respect to F^x, $x \in \Omega$, it is necessary and sufficient that M^x commutes with C.

Proof. Necessity: Let

$$\bar{P}^x = (\ldots, P^{y'}, \ldots)', \tag{4.4.1}$$

where P^y is included in $\bar{P}^x$ for every $y \in \Omega$, $y \neq x$ (for example, if $n = 2$, $x = 01$, then $\bar{P}^x = (P^{10'}, P^{11'})'$). Let the design have POFS with respect to F^x. Then $Cov(P^x \hat{\underline{\tau}}, \bar{P}^x \hat{\underline{\tau}}) = 0$, and proceeding as in the 'necessity' part of Theorem 4.2.1,

$$C = P^{x'} P^x C P^{x'} P^x + \bar{P}^{x'} \bar{P}^x C \bar{P}^{x'} \bar{P}^x. \tag{4.4.2}$$

Hence by (2.2.7), (2.2.8), (4.4.1), $M^x C = M^x C M^x = C M^x$.

Sufficiency: Let M^x commute with C. Then by (4.4.1) and Lemma 3.1.2, $\bar{P}^{x'} \bar{P}^x$ also commutes with C. Hence as in 'sufficiency' part of Theorem 4.2.1, the BLUE's of $P^x \underline{\tau}$ and $\bar{P}^x \underline{\tau}$ are given by

$$P^x \hat{\underline{\tau}} = (P^x C P^{x'})^{-1} P^x \underline{Q}, \quad \bar{P}^x \hat{\underline{\tau}} = (\bar{P}^x C \bar{P}^{x'})^{-1} \bar{P}^x \underline{Q},$$

and, exactly as in that theorem, it follows that $Cov(P^x \hat{\underline{\tau}}, \bar{P}^x \hat{\underline{\tau}}) = 0$, so that the design has POFS with respect to F^x.

In dealing with a design which possibly does not have OFS, the above theorem is of much help in identifying the pair(s) of interactions which are not mutually orthogonally estimated. The following example illustrates the point.

Example 4.4.1 (Chauhan and Dean, 1986). Consider a $4 \times 2 \times 2$ factorial in 16 blocks each of size 6, by adding in turn the treatment combinations (000, 100, 200, 300) to the following four blocks

{000 001 100 101 210 311},
{001 010 101 110 211 300},
{010 011 110 111 200 301},
{011 000 111 100 201 310},

where addition of two treatment combinations $(j_1j_2j_3)$ and $(j_1'j_2'j_3')$ is defined by $(j_1^*j_2^*j_3^*) = (j_1j_2j_3) + (j_1'j_2'j_3')$, with $j_i^* = j_i + j_i'$ mod m_i, $i = 1, 2, 3$. For this design,

$$NN' = \begin{bmatrix} E_1 & E_2 & E_3 & E_4 \\ E_4 & E_1 & E_2 & E_3 \\ E_3 & E_4 & E_1 & E_2 \\ E_2 & E_3 & E_4 & E_1 \end{bmatrix},$$

where E_1, E_2, E_3, E_4 are circulant matrices, their initial rows being given by (6 2 0 2), (2 4 2 1), (0 2 4 2), (2 1 2 4) respectively. It may be checked that M^x commutes with NN', and hence with C, for every x other than $x = 110, 111$. Hence POFS holds with respect to all interactions other than F^{110} and F^{111}. In other words, all pairs of interactions are orthogonal with the exception of (F^{110}, F^{111}).

The following corollary is a direct consequence of (4.4.2).

Corollary 4.4.1 (Chauhan and Dean, 1986). In a connected factorial design if POFS holds with respect to F^x, $x \in \Omega$, then the C-matrix can be expressed as

$$C = C^x + \overline{C}^x,$$

where the matrices C^x, $\overline{C}^x$ are symmetric and $R(C^x) \subset R(P^x)$, $R(\overline{C}^x) \subset R(\overline{P}^x)$.

We now consider the problem of orthogonal estimation of contrasts belonging to lower order interactions in general. To that effect, an n-factor design will be said to have POFS of order $t(\leq n)$ if the design has POFS with respect to every interaction F^x involving at most t factors. Let Ω_t^* be the set of all n-component non-null binary vectors having at most t components equal to unity (for example, if $n = 3$, then $\Omega_2^* =$ {001,010,100,011,101,110}). Obviously, interactions involving at most t factors are given by F^x, $x \in \Omega_t^*$. Hence the following result is evident from Theorem 4.4.1.

Theorem 4.4.2. For a connected factorial design to have POFS of order t, it is necessary and sufficient that M^x commutes with C for every $x \in \Omega_t^*$.

Mukerjee (1980) obtained another characterization for POFS of order t. For $t \leq n$, let Ω_t be the set of all n-component non-null binary vectors having exactly t components equal to unity (for example, if n = 3, then $\Omega_2 = \{011,101,110\}$). Then the following result holds.

Theorem 4.4.3. For a connected factorial design to have POFS of order t, it is necessary and sufficient that for every $x \in \Omega_t$, (i) Z^x commutes with C and (ii) $Z^x C$ has structure K.

The proof of Theorem 4.4.3 is omitted here and the interested reader may refer to Mukerjee (1980) for the same. Analogously to Theorem 4.2.8, it is possible to derive yet another characterization.

Theorem 4.4.4. For a connected factorial design to have POFS of order t, it is necessary and sufficient that for every $x \in \Omega_t^*$, the C-matrix has $\alpha(x) = \Pi(m_i-1)^{x_i}$ orthonormal eigenvectors belonging to $\mu(P^{x'})$.

The characterizations for POFS considered so far are all directly in terms of the C-matrix of the design. Chauhan and Dean (1986) obtained an important characterization in terms of a g-inverse of the C-matrix. The result is of considerable help in verifying whether or not contrasts belonging to a given pair of interactions are orthogonally estimated. It is based on the following lemma.

Lemma 4.4.1 (Chauhan and Dean, 1986). Let B_1, B_2, G be non-null real matrices such that the product B_1GB_2 exists. Then $B_1GB_2 = 0$ if and only if $G = G_1 + G_2$ where $B_1G_1 = 0$, $G_2B_2 = 0$.

Proof. The 'if' part is obvious. To prove the 'only if' part, observe that if $B_1GB_2 = 0$, then from Rao and Mitra (1971, Theorem 2.3.2)

$$\begin{aligned} G &= G_0 - B_1^-B_1G_0B_2B_2^- \\ &= (I-B_1^-B_1)G_0 + B_1^-B_1G_0(I-B_2B_2^-) = G_1 + G_2, \end{aligned}$$

say, where G_0 is an arbitrary matrix and B_1^-, B_2^- are g-inverses of B_1, B_2 respectively. The matrices G_1, G_2 clearly satisfy the statement of the lemma .

In a connected factorial design, for $x \in \Omega$, the BLUE of $P^x \underline{\tau}$ is given by $P^x \hat{\underline{\tau}} = P^x \overline{C} \underline{Q}$, as before $\overline{C}$ being any g-inverse of the C-matrix. Hence for $x, y \in \Omega$, $x \neq y$ $Cov(P^x \hat{\underline{\tau}}, P^y \hat{\underline{\tau}}) = \sigma^2 P^x \overline{C} P^{y'}$. The following result is now an immediate consequence of Lemma 4.4.1.

Theorem 4.4.5 (Chauhan and Dean, 1986). In a connected factorial design, the relation $Cov(P^x \hat{\underline{\tau}}, P^y \hat{\underline{\tau}}) = 0$, holds for a fixed pair $x, y \in \Omega$, $x \neq y$, if and only if any g-inverse $\overline{C}$ of the C-matrix can be expressed as $\overline{C} = \overline{C}^x + \overline{C}^y$, where $P^x \overline{C}^x = 0$, $\overline{C}^y P^{y'} = 0$.

For interesting corollaries to the above theorem, we refer to Chauhan and Dean (1986). Some further results, useful in the construction of designs with POFS, were also obtained by Chauhan and Dean (1986). These aspects will be considered later in this monograph. The following theorem gives a characterization for POFS in the disconnected case.

Theorem 4.4.6 (Chauhan and Dean, 1986). For a regular disconnected factorial design to have POFS with respect to F^x, $x \in \Omega$, it is necessary and sufficient that M^x commutes with C.

Proof. Necessity: With the notation as in the preceding section, for each $y \in \Omega$, let H_y be a matrix whose rows form an orthonormal basis of V_y^*. Let

$$H_x^* = (\ldots, H_y', \ldots)', \tag{4.4.3}$$

where H_y is included in H_x^* for every $y \in \Omega$, $y \neq x$ (cf. (4.4.1)). If a regular disconnected design has POFS with respect to F^x, then exactly as in the proof of Theorem 4.3.1, one obtains

$$C = H_x' H_x C H_x' H_x + H_x^{*\prime} H_x^* C H_x^{*\prime} H_x^*.$$

Hence by (2.2.7), (2.2.8), (4.3.2), (4.4.3), it follows that $M^x C = H_x' H_x C H_x' H_x = C M^x$, since $M^x H_x' H_x = H_x' H_x M^x = H_x' H_x$ and $M^x H_x^{*\prime} H_x^* = H_x^{*\prime} H_x^* M^x = 0$.

Sufficiency: Let $\overline{P}^x$ be as in (4.4.1) and $\overline{H}_x$ be a matrix whose rows form an orthonormal basis of $R(\overline{P}^x) \cap R(C)$. If M^x commutes with C, then as in the proof of Theorem 4.4.1, $\overline{P}^{x\prime}\overline{P}^x$ also commutes with C and, analogously to (4.3.9), the BLUE's of $H_x\underline{\tau}$ and $\overline{H}_x\underline{\tau}$ are given by

$$H_x\hat{\underline{\tau}} = U_x M^x \underline{Q}, \; \overline{H}_x\hat{\underline{\tau}} = \overline{U}_x \overline{P}^{x\prime}\overline{P}^x \underline{Q},$$

respectively, where the matrices U_x, $\overline{U}_x$ are such that $H_x = U_x C$, $\overline{H}_x = \overline{U}_x C$. Under commutativity of M^x and C, it now follows, as in the 'sufficiency' part of Theorem 4.2.1, that $Cov(H_x\hat{\underline{\tau}}, \overline{H}_x\hat{\underline{\tau}}) = 0$.

It may be remarked that the proof of the 'sufficiency' part of Theorem 4.4.6 does not require the information that the disconnected design under consideration is regular. It is now easy to extend Theorems 4.4.2, 4.4.3, 4.4.4 to the disconnected case. In fact, it is readily seen that all these theorems remain valid if the word 'connected' in their statements be replaced by the phrase 'regular disconnected'.

Chauhan (1988) considered the notion of partial regularity with respect to a specific interaction. With $\overline{P}^x$ given by (4.4.1), let $\overline{V}_x^* = R(\overline{P}^x) \cap R(C)$. In analogy with Definition 4.3.1, a design is partially regular with respect to the interaction F^x, $x \in \Omega$, if $V_x^* \oplus \overline{V}_x^*$ equals $R(C)$. Specializing to single replicate designs, Chauhan (1987, 1988) proved that a design is partially regular with respect to F^x if and only if M^x commutes with C and derived some further characterizations. For the details, we refer to the original papers.

4.5. Efficiency-consistency

In the preceding sections, we have considered several algebraic characterizations for OFS and POFS. The present section deals with another type of characterization which is helpful in demonstrating how OFS or POFS may lead to an elegant interpretation of interaction efficiencies in a factorial design.

Consider an equireplicate factorial design, d, in n factors such that each treatment combination is replicated r times in d. For each $x = (x_1, \ldots, x_n) \in \Omega$, let d_x denote the design formed from d by deleting the ith digit from the treatment labels for all i for which $x_i = 0$; $i = 1,\ldots,n$. Thus with reference to the 3×4 factorial considered in Example 3.1.1, the design d_{01} is as shown below:

<---- Blocks ---->

0	0	0	1	1	1	2	2	2	3	3	3
1	2	3	0	2	3	0	1	3	0	1	2
2	3	1	3	0	2	1	3	0	2	0	1

Definition 4.5.1 (Lewis and Dean, 1985). An equireplicate n-factor design d is efficiency-consistent if for every $x \in \Omega$, the efficiencies of all estimable contrasts belonging to F^x are equal to the efficiencies of the equivalent contrasts in d_x.

The main result of this section is Theorem 4.5.1 which establishes the equivalence of efficiency-consistency and OFS for equireplicate, connected factorial designs. The 'sufficiency' part of this theorem was proved by Lewis and Dean (1985) while the 'necessity' part was established by Mukerjee and Dean (1986). The following considerations will be helpful in proving the theorem.

Defining $\bar{P}^x$ as in (4.4.1), it is easy to see that for a connected factorial design d, which does not necessarily have OFS, the dispersion matrix of $P^x \hat{\underset{\sim}{\tau}}$, $x \in \Omega$, is given by $\sigma^2 W_{1x}$, where

$$W_{1x} = [P^x C P^{x'} - P^x C \bar{P}^{x'} (\bar{P}^x C \bar{P}^{x'})^{-1} \bar{P}^x C P^{x'}]^{-1}, \tag{4.5.1}$$

σ^2 is the constant error variance and, as usual, C is the C-matrix of the design. Note that the matrix inverses in (4.5.1) are well-defined since by Lemma 3.1.3, for a connected design PCP' is p.d., where $P = (P^{x'}, \bar{P}^{x'})'$.

For a fixed $x \in \Omega$, consider the subdesign d_x. Let $v_x = \Pi m_i^{x_i}$, and $\underset{\sim}{\tau}_x$ denote the $v_x \times 1$ vector of factorial treatment effects in d_x. The C-matrix of d_x is given by, say,

$$C_x = \epsilon^x C \epsilon^{x'}, \tag{4.5.2}$$

where

$$\epsilon^x = \bigotimes_{i=1}^{n} \epsilon_i^{x_i}, \tag{4.5.3}$$

and for $1 \le i \le n$,

$$\left.\begin{aligned} \epsilon_i^{x_i} &= I_i \quad \text{if } x_i = 1 \\ &= \underline{1}_i' \quad \text{if } x_i = 0 \end{aligned}\right\} \tag{4.5.4}$$

Let $\Omega(x) = \{y:\ y \in \Omega,\ y_i \le x_i,\ \forall i\}$; for example, if $n = 3$ then $\Omega(011) = \{001,010,011\}$. Defining

$$B_x^y = (v_x/v)^{1/2} P^y\, \epsilon^{x'}, \tag{4.5.5}$$

it follows that $B_x^y \underline{\tau}_x$ represents a complete set of orthonormal treatment contrasts belonging to F^y in d_x, for each $y \in \Omega(x)$. By (2.2.5), (2.2.6), (4.5.3), (4.5.4), for $y \in \Omega(x)$,

$$P^y \epsilon^{x'} \epsilon^x = (v/v_x) P^y$$

and hence by (4.5.2) - (4.5.5), for $y,\ z \in \Omega(x)$

$$B_x^y C_x B_x^{z'} = (v/v_x) P^y C P^{z'}. \tag{4.5.6}$$

The connectedness of d implies that of d_x. Let $B_x^x\hat{\underline{\tau}}_x$ denote the BLUE of $B_x^x \underline{\tau}_x$ in d_x. Then by (4.5.6), the dispersion matrix of $B_x^x\hat{\underline{\tau}}_x$ is given by $\sigma^2 W_{2x}$, where analogously to (4.5.1),

$$W_{2x} = (v_x/v)(P^x C P^{x'})^{-1}, \text{ if } F^x \text{ represents a main effect} \tag{4.5.7a}$$

$$= (v_x/v)[P^x C P^{x'} - P^x C \bar{P}_x^{x'} (\bar{P}_x^x C \bar{P}_x^{x'})^{-1} \bar{P}_x^x C P^{x'}]^{-1}, \text{ otherwise.} \tag{4.5.7b}$$

In the above,

$$\bar{P}_x^x = (\ldots, P^{y'}, \ldots)',$$

where P^y is included in $\bar{P}_x^x$ for every $y \in \Omega(x),\ y \ne x$. The fact that $\Omega(x) = \{x\}$ when F^x represents a main effect explains (4.5.7a). The connectedness of d, and hence that of d_x, guarantees the existence of the matrix inverses in (4.5.7a, b).

Since the design d is equireplicate with a common replication number r, the treatment combinations in d_x are each replicated $r_x = (rv/v_x)$ times. The BLUE of a typical contrast $\underline{u}' P^x \underline{\tau}\ (\underline{u} \ne \underline{0})$ has efficiency $r^{-1}\underline{u}'\underline{u}/(\underline{u}' W_{1x} \underline{u})$ in d, while that of the equivalent contrast $\underline{u}' B_x^x \underline{\tau}_x$ has efficiency $r_x^{-1}\underline{u}'\underline{u}/(\underline{u}' W_{2x}\underline{u})$ in d_x. Hence by Definition 4.5.1, the design d is efficiency-consistent if and only if

$$r^{-1}\underline{u}'\underline{u}/(\underline{u}'W_{1x}\underline{u}) = r_x^{-1}\underline{u}'\underline{u}/(\underline{u}'W_{2x}\underline{u}),$$

for all $\underline{u} \neq \underline{0}$ and for all $x \in \Omega$, that is if and only if $W_{1x} = (v/v_x)W_{2x}$ for all $x \in \Omega$. Defining the sets $\Omega_t (1 \leq t \leq n)$ as in the preceding section, it follows from (4.5.1) and (4.5.7a,b) that d is efficiency-consistent if and only if

$$P^x C\overline{P}^{x\prime}(\overline{P}^x C\overline{P}^{x\prime})^{-1}\overline{P}^x CP^{x\prime} = 0, \qquad \text{for } x \in \Omega_1, \qquad (4.5.8a)$$

$$= P^x C\overline{P}_x^{x\prime}(\overline{P}_x^x C\overline{P}_x^{x\prime})^{-1}\overline{P}_x^x CP^{x\prime}, \text{ for } x \in \Omega - \Omega_1 . \qquad (4.5.8b)$$

For brevity, the left- and right-hand members of (4.5.8a,b) will be denoted by L and R respectively.

Theorem 4.5.1. For an equireplicate, connected factorial design to be efficiency-consistent, it is necessary and sufficient that the design has OFS.

Proof. Sufficiency: If the design has OFS, then by Theorem 4.2.6, $P^x C\overline{P}^{x\prime} = 0$ and $P^x C\overline{P}_x^{x\prime} = 0$ for every $x \in \Omega$. Thus $L = R = 0$ for all $x \in \Omega$ and (4.5.8a,b) hold.

Necessity: Let the design d be efficiency-consistent so that (4.5.8a, b) hold. If $x \in \Omega_1$, then by (4.5.8a), $L = 0$. Hence $P^x C\overline{P}^{x\prime} = 0$ and it follows that

$$P^x CP^{y\prime} = 0, \ \forall y \in \Omega, \ y \neq x. \qquad (4.5.9)$$

We prove that (4.5.9) holds for all $x \in \Omega$ by induction. Assume that (4.5.9) holds for $x \in \Omega_1, \Omega_2, \ldots, \Omega_g$ $(1 \leq g < n)$ and consider $x \in \Omega_{g+1}$. If $y \in \Omega(x)$ and $y \neq x$ then $y \in \bigcup_{i=1}^{g} \Omega_i$. Reversing the roles of x and y in (4.5.9), it follows that $P^x C\overline{P}_x^{x\prime} = 0$. Hence in (4.5.8b), L = R = 0 and, as such, (4.5.9) holds for $x \in \Omega_{g+1}$. Thus by induction (4.5.9) holds and from Theorem 4.2.6, the design has OFS.

Theorem 4.5.1 may be extended in a routine manner to regular disconnected designs replacing, throughout the derivation, the matrices P^x and $\overline{P}^x$ respectively by H_x and H_x^* which are as in the proof of Theorem 4.4.6 - see Mukerjee and Dean (1986) for the details. This leads to the following result.

Theorem 4.5.2. For an equireplicate, regular, disconnected factorial design to be efficiency-consistent, it is necessary and sufficient that the design has OFS.

Results on the connection between various types of POFS and 'partial' efficiency-consistency are also available.

Definition 4.5.2. (i) An equireplicate n-factor design d is partially efficiency - consistent with respect to F^x for a fixed $x \in \Omega$, if the efficiencies of all estimable constrasts belonging to F^x are equal to the efficiencies of the equivalent contrasts in d_x. (ii) The design d is partially efficiency-consistent of order $t(\leq n)$ if it is partially efficiency-consistent with respect to every interaction F^x involving at most t factors.

Theorem 4.5.3 (Mukerjee and Dean, 1986). For an equireplicate factorial design, which is either connected or regular disconnected, to be partially efficiency-consistent of order $t(\leq n)$, it is necessary and sufficient that the design has POFS of order t.

Theorem 4.5.4 (Gupta, 1986b). If an equireplicate factorial design, which is either connected or regular disconnected, has POFS with respect to F^x for a fixed $x \in \Omega$, then the design is partially efficiency-consistent with respect to F^x.

The proofs of Theorems 4.5.3, 4.5.4 follow essentially along the line of proof of Theorem 4.5.1 and are hence omitted here. It may be remarked that the converse of Theorem 4.5.4 is not necessarily true, i.e., partial efficiency-consistency with respect to a particular F^x may not imply POFS with respect to F^x. In order to appreciate this point, it is enough to observe that an n-factor equireplicate design is always partially efficiency-consistent with respect to the n-factor interaction (for then d and d_x become identical), but this does not necessarily guarantee POFS with respect to this interaction. The following definition is helpful in obtaining a necessary and sufficient condition for partial efficiency-consistency with respect to a particular interaction.

Definition 4.5.3. For a fixed $x \in \Omega$, a factorial design will be said to have (i) external POFS with respect to F^x if the BLUE's of estimable contrasts belonging to F^x are uncorrelated with the BLUE's of estimable contrasts belonging to F^y for every

$y \in \Omega - \Omega(x)$, (ii) internal POFS with respect to F^x if the BLUE's of estimable contrasts belonging to F^x are uncorrelated with the BLUE's of estimable contrasts belonging to F^y for every $y \in \Omega(x)$, $y \neq x$.

Clearly, a design has POFS with respect to F^x in the sense of Definition 4.4.1 if and only if it has both external and internal POFS with respect to F^x.

Theorem 4.5.5. (Mukerjee and Dean, 1986). For an equireplicate factorial design, which is either connected or regular disconnected, to be partially efficiency-consistent with respect to F^x for a fixed $x \in \Omega$, it is necessary and sufficient that the design has external POFS with respect to F^x.

The proof of Theorem 4.5.5 may be worked out along the line of proof of Theorem 4.5.1.

Theorems 4.5.1 - 4.5.5 demonstrate the connection between OFS (POFS) and efficiency-consistency (partial efficiency-consistency). It may be of some interest to consider a similar characterization for the phenomenon of regularity. To that effect, an n-factor design d will be defined as estimability-consistent provided for each $x \in \Omega$, each contrast belonging to F^x in d is estimable in d if and only if the equivalent contrast belonging to F^x in d_x is estimable in d_x. Recently, it was proved by Mukerjee and Bose (1988a) that a factorial design is regular if and only if it is estimability-consistent. This finding provides, in a sense, a simple and intuitively appealing interpretation for the somewhat abstract phenomenon of regularity. The proof of the result is omitted here.

We conclude this chapter with the remark that all the theorems presented here, except Theorem 4.2.2, extend themselves in a natural manner to row-column designs or designs for multiway heterogeneity elimination provided C is interpreted as the coefficient matrix of the reduced normal equations for treatment effects in such designs.

CHAPTER 5

CONSTRUCTIONS I: FACTORIAL EXPERIMENTS IN CYCLIC AND GENERALIZED CYCLIC DESIGNS

5.1. Introduction

In recent years two broad methods emerged for the construction of factorial designs: (a) use of cyclic or generalized cyclic designs and (b) use of Kronecker or Kronecker-type products. Both the methods lead to designs with OFS and, if appropriately used, are capable of ensuring high efficiencies with respect to the interactions of interest. In this chapter, the method (a) will be reviewed. The method (b) will be taken up in the next chapter.

5.2. Factorial experiments in cyclic designs

Cyclic designs are incomplete block designs consisting in the simplest case of a set of blocks obtained by the development of an initial block. More generally, a cyclic design consists of a combination of such sets. A cyclic design can, therefore, be specified by its initial block or blocks. For the main results on cyclic designs, we refer to David and Wolock (1965) and John (1966, 1987). A catalogue of cyclic designs is given by John, Wolock and David (1972). Following John (1973a), we consider in this section the use of cyclic designs for two-factor experiments. The following lemma will be helpful.

Lemma 5.2.1. For an $m_1 \times m_2$ factorial design, let the C-matrix be of the form

$$C = \begin{bmatrix} C_0 & C_1 & \cdots & C_{m_1-1} \\ C_{m_1-1} & C_0 & \cdots & C_{m_1-2} \\ \cdot & \cdot & \cdots & \cdot \\ \cdot & \cdot & \cdots & \cdot \\ \cdot & \cdot & \cdots & \cdot \\ C_1 & C_2 & \cdots & C_0 \end{bmatrix} \tag{5.2.1}$$

where each C_j is a square matrix of order m_2. Then in order that C may have structure K, it is necessary and sufficient that for each $j\,(0 \leq j \leq m_1-1)$, C_j is proper.

Proof. Necessity: Let C have structure K. Then by Theorem 4.2.7, $Z^{10}(=I_1 \otimes J_2)$ commutes with C. Hence for each j, J_2 commutes with C_j so that C_j is proper $(0 \leq j \leq m_1-1)$.

Sufficiency: For $0 \leq j \leq m_1 - 1$, let R_{1j}^* be $m_1 \times m_1$ circulant matrices as in Theorem 4.2.9. Then

$$C = \sum_{j=0}^{m_1-1} R_{1j}^* \otimes C_j. \tag{5.2.2}$$

Since $R_{1j}^* \; (0 \leq j \leq m_1 - 1)$ are proper matrices, the sufficiency of the stated condition follows immediately.

Consider now an $m_1 \times m_2$ factorial experiment in a cyclic design. Since the design is cyclic, its C-matrix can be expressed as a linear combination of circulant permutation matrices of order v $(=m_1 m_2)$. Consequently, C must be of the form (5.2.1). Moreover, for each j, C_j can be written as

$$C_j = \begin{bmatrix} c_0^{(j)} & c_1^{(j)} & \cdots & c_{m_2-1}^{(j)} \\ c_{m_2-1}^{(j-1)} & c_0^{(j)} & \cdots & c_{m_2-2}^{(j)} \\ \cdot & \cdot & \cdots & \cdot \\ \cdot & \cdot & \cdots & \cdot \\ \cdot & \cdot & \cdots & \cdot \\ c_2^{(j-1)} & c_3^{(j-1)} & \cdots & c_1^{(j)} \\ c_1^{(j-1)} & c_2^{(j-1)} & \cdots & c_0^{(j)} \end{bmatrix},$$

where (j-1) is reduced modulo m_1. Then the following theorem holds:

Theorem 5.2.1. In order that the C-matrix may have structure K, it is necessary and sufficient that for each j $(0 \leq j \leq m_1-1)$, C_j is symmetric and circulant and that, apart from the diagonal elements, all submatrices in C are equal.

Proof. Necessity: Let C have structure K. Then by Lemma 5.2.1, each C_j is proper. Hence, considering pairs of consecutive rows in each C_j,

$$c_u^{(j)} = c_u^{(j-1)}, \quad 1 \le u \le m_2-1, \quad 0 \le j \le m_1 - 1.$$

This implies that $c_u^{(0)} = c_u^{(1)} = \cdots = c_u^{(m_1-1)}$ $(1 \le u \le m_2-1)$. It follows that for each j, C_j is circulant and that, apart from the diagonal elements, all submatrices in C are equal. The symmetry of each C_j now follows from that of C_0.

Sufficiency: Follows trivially from Lemma 5.2.1.

In view of Theorems 4.2.1, 4.3.1 and 4.3.2, the condition in Theorem 5.2.1 characterize OFS (together with regularity, in the disconnected case) for two-factor experiments in cyclic designs. It may be remarked that John (1973a) considered a version of Theorem 5.2.1 in terms of a g-inverse of the C-matrix. If the conditions in Theorem 5.2.1 are satisfied then the analysis of the design may be performed using (4.2.11) or (4.3.11).

Example 5.2.1. (John, 1973a). Consider a cyclic design in 12 treatments and 24 blocks obtained by the development of the initial blocks (0,1,4) and (0,2,5). This may be looked upon as a 4×3 factorial design by identifying the treatment $4j_1 + j_2$ with the treatment combination (j_1, j_2), $0 \le j_1 \le 3$, $0 \le j_2 \le 2$. It may then be seen that the C-matrix of the design is of the form (5.2.1) with $m_1 = 4$ and

$$C_0 = \frac{1}{3}\begin{pmatrix} 12 & -1 & -1 \\ -1 & 12 & -1 \\ -1 & -1 & -1 \end{pmatrix} \quad C_1 = -\frac{1}{3}\begin{pmatrix} 2 & 1 & 1 \\ 1 & 2 & 1 \\ 1 & 1 & 2 \end{pmatrix},$$

$$C_2 = -\frac{1}{3}\begin{pmatrix} 0 & 1 & 1 \\ 1 & 0 & 1 \\ 1 & 1 & 0 \end{pmatrix}, \quad C_3 = -\frac{1}{3}\begin{pmatrix} 2 & 1 & 1 \\ 1 & 2 & 1 \\ 1 & 1 & 2 \end{pmatrix}.$$

Clearly, the conditions of Theorem 5.2.1 are satisfied so that the design has OFS.

It is interesting to note that, whether the condition in Theorem 5.2.1 are satisfied or not, a two-factor experiment in a cyclic design always possesses POFS with respect to one main effect. This is shown in the following theorem.

Theorem 5.2.2. An $m_1 \times m_2$ factorial experiment in a cyclic design has POFS with respect to the main effect of the second factor.

Proof. Note that the C-matrix of the design is of the form (5.2.2). Hence with M^{01} defined as in (2.2.1),

$$M^{01}C = (m_1^{-1}J_1) \otimes \{(I_2 - m_2^{-1}J_2)(\sum_{j=0}^{m_1-1} C_j)\}. \tag{5.2.3}$$

Since all the column sums of C sum to zero, i.e., $(J_1 \otimes J_2)C = 0$, it follows from (5.2.2.) that $J_1 \otimes \{J_2(\sum_{j=0}^{m_1-1} C_j)\} = 0$. Hence by (5.2.3),

$$M^{01}C = (m_1^{-1}J_1) \otimes (\sum_{j=0}^{m_1-1} C_j). \tag{5.2.4}$$

Since C is symmetric, one has $C_j' = C_{m_1-j}(0 \leq j \leq m_1-1)$, where $m_1 - j$ is reduced modulo m_1. Consequently, $\sum_{j=0}^{m_1-1} C_j$ is symmetric and, by (5.2.4.), M^{01} commutes with C. The result now follows from Theorem 4.4.1.

For some results on factorial experiments in four-associate class cyclic partially balanced designs, we refer to Gupta (1986c).

5.3. Generalized cyclic designs

John (1973b) introduced generalized cyclic designs in the context of factorial experiments. Consider an n-factor setting with factors at $m_1, \ldots, m_n$ (≥ 2) levels respectively. A generalized cyclic set is generated from an initial block consisting of k treatment combinations. For each treatment combination $(j_1, \ldots, j_n)$ $(\neq (0, \ldots, 0))$ a block is generated by adding $(j_1, \ldots, j_n)$ to each treatment combination in the initial block; here addition of two treatment combinations $(j_1, \ldots, j_n)$ and $(j_1', \ldots, j_n')$ is defined by

$$(j_1^*, \ldots, j_n^*) = (j_1, \ldots, j_n) + (j_1', \ldots, j_n'),$$

with $j_i^* = j_i + j_i'$ mod m_i $(1 \leq i \leq n)$. Under this definition a set will consist of

$v = \Pi m_i$ blocks.

Example 5.3.1. Let $n = 2$, $m_1 = 4$, $m_2 = 3$. The generalized cyclic set obtained from the initial block {00,01,02,10} is given by

{00,01,02,10}	{01,02,00,11}	{02,00,01,12}
{10,11,12,20}	{11,12,10,21}	{12,10,11,22}
{20,21,22,30}	{21,22,20,31}	{22,20,21,32}
{30,31,32,00}	{31,32,30,01}	{32,30,31,02}

It can be seen that successive blocks in any column are generated from the first block of that column by cycling the first digit of a treatment combination; similarly for rows the second digit is cycled.

Some generalized cyclic sets will have a fraction, 1/p say, of the v blocks replicated p times. Such sets will be called fractional sets and will contain v/p distinct blocks. For example, with $m_1 = 4$, $m_2 = 3$, a fractional set of 3 distinct blocks can be generated from the initial block {00,10,20,30}.

Generalized cyclic (GC/n) designs are constructed by taking full or fractional sets individually or in combination, and will be of size $(v,k,r;m_1, \ldots, m_n)$, where k and r represent the constant block size and the common replication number respectively. For such designs, it can be seen that NN' is block circulant, i.e.,

$$NN' = \sum_{j_1=0}^{m_1-1} \cdots \sum_{j_n=0}^{m_n-1} \xi_{j_1 \cdots j_n} \left(\bigotimes_{i=1}^{n} R^*_{ij_i} \right), \tag{5.3.1}$$

when $\xi_{j_1 \cdots j_n}$ are scalars and $R^*_{ij_i}$ are circulant permutation matrices as defined in Section 4.2 $(0 \le j_i \le m_i - 1,\ 1 \le i \le n)$. Clearly, NN' has structure K. Since GC/n designs are equireplicate and proper, it follows from Theorems 4.2.2, 4.3.2 that such designs have OFS (and are regular, if disconnected). Consequently, one may employ (4.2.11) or (4.3.11) to obtain the sums of squares in such designs.

In order to derive expressions for interaction efficiencies in GC/n designs, observe from (5.3.1) that the elements in the initial row of NN' are given by the $\xi_{j_1 \cdots j_n}$, arranged lexicographically. Since NN' is symmetric, the relation

$$\xi_{j_1 \cdots j_n} = \xi_{j_1' \cdots j_n'} \tag{5.3.2}$$

holds for each $j_1, \ldots, j_n$, where $j_i' = m_i - j_i \mod m_i (1 \le i \le n)$. Recall that by Lemma 2.2.3, for each $x \in \Omega$, $P^x \underline{\tau}$ represents a complete set of orthonormal contrasts belonging to the interaction F^x. Since for a GC/n design, the matrix NN', and hence C, has structure K, it is clear from the results in Chapter 4 that the information matrix for $P^x \underline{\tau}$ is given by $P^x C P^{x'}$ (cf. (4.2.11) or (4.3.11)). Let $\lambda_1^x, \ldots, \lambda_{\alpha(x)}^x$ be the eigenvalues of $P^x C P^{x'}$, where $\alpha(x) = \Pi(m_i - 1)^{x_i}$. Then the efficiency of a GC/n design with respect to the interaction F^x is defined as

$$\epsilon(x) = \left.\begin{matrix} r^{-1}\alpha(x)\{\sum_{j=1}^{\alpha(x)}(\lambda_j^x)^{-1}\}^{-1} & \text{if } \lambda_j^x > 0 \ (1 \le j \le \alpha(x)) \\ 0 & otherwise \end{matrix}\right\}, \tag{5.3.3}$$

where r is the number of replications in the design. As usual, here efficiency is relative to a randomized (complete) block design having the same number of replicates. In fact, (5.3.3) defines A-efficiency (cf. Kiefer (1975)) as used by John (1973a). A more general definition of efficiency will be considered in the next chapter.

With GC/n designs, it is possible to work out a further reduction of (5.3.3) in terms of trigonometric functions. To that effect, for the sake of notational simplicity consider, without loss of generality, an interaction involving the first g factors $(1 \le g \le n)$, i.e., consider F^x, where $x = (x_1, \ldots, x_n)$, with $x_1 = \cdots = x_g = 1$, $x_{g+1} = \cdots = x_n = 0$. Then by (2.2.5), (2.2.6), (5.3.1),

$$P^x NN' P^{x'} = \sum_{j_1=0}^{m_1-1} \cdots \sum_{j_n=0}^{m_n-1} \xi_{j_1 \cdots j_n} (\bigotimes_{u=1}^{g} P_u R_{uj_u}^{*} P_u'). \tag{5.3.4}$$

Let $\psi_{u0} = 1, \psi_{u1}, \ldots, \psi_{um_u-1}$ be the m_u-th roots of unity, i.e., $\psi_{uh} = exp(2\pi ih/m_u)$, $0 \le h \le m_u - 1$, $1 \le u \le n$, where $i^2 = -1$. Let $\gamma_{uh} = (1, \psi_{uh}, \ldots, \psi_{uh}^{m_u-1})'$, $0 \le h \le m_u - 1$, $1 \le u \le n$. Then it may be seen that for each $h_1, \ldots, h_g$ $(1 \le h_u \le m_u - 1, 1 \le u \le g)$,

$$(\bigotimes_{u=1}^{g} P_u)(\bigotimes_{u=1}^{g} \gamma_{uh_u})$$

is an eigenvector of $(\bigotimes_{u=1}^{g} P_u R_{uj_u}^{*} P_u')$, the corresponding eigenvalue being given by

$\prod_{u=1}^{g} \psi_{uh_u}^{j_u}$. Hence by (5.3.4), the eigenvalues of $P^x NN' P^{x'}$ are given by

$$\sum_{j_1=0}^{m_1-1} \cdots \sum_{j_n=0}^{m_n-1} \xi_{j_1 \ldots j_n} \exp(2\pi i \sum_{u=1}^{g} h_u j_u / m_u)$$
$$= \sum_{j_1=0}^{m_1-1} \cdots \sum_{j_n=0}^{m_n-1} \xi_{j_1 \ldots j_n} \cos(2\pi \sum_{u=1}^{g} h_u j_u / m_u), \quad 1 \le h_u \le m_u - 1, \; 1 \le u \le g, \tag{5.3.5}$$

after some simplification using (5.3.2) - for details, see Williams (1975) as reported in John and Lewis (1983). Since $C = rI - k^{-1}NN'$, it is now easy to work out the eigenvalues of $P^x CP^{x'}$ and then calculate $\epsilon(x)$ employing (5.3.3). The efficiencies for other interactions may be obtained in a similar manner.

The relation (5.3.5) reveals an interesting feature of GC/n designs. The computation of interaction efficiencies in such designs is rather straightforward in the sense that one has to consider only the initial row of NN'. Consequently, for given $m_1, \ldots, m_n, r, k$, it is possible to compute quickly the interaction efficiencies in the available GC/n designs and then to select the one with high efficiencies with respect to the interactions of interest.

Example 5.3.2. (John, 1973b). With $n = 2$, $m_1 = 4$, $m_2 = 3$, $r = k = 4$, consider the $GC/2$ design shown in Example 5.3.1. For this design, the initial row of NN' is given by (4 3 3 1 1 1 0 0 0 1 1 1). Hence by (5.3.5), it may be seen that the eigenvalues of $P^{10}CP^{10'}$ are 1.5, 1.5 and 3. Similarly, $P^{01}CP^{01'}$ has an eigenvalue 3.75 with multiplicity 2 and $P^{11}CP^{11'}$ has an eigenvalue 3.75 with multiplicity 6. Hence by (5.3.3.), for this design $\epsilon(10) = 0.4500$, $\epsilon(01) = \epsilon(11) = 0.9375$. In a similar manner, it may be seen that for the GC/2 design with the same parameters, generated from the initial block {00, 10, 22, 31}, the efficiencies are given by $\epsilon(10) = 1.0000$, $\epsilon(01) = 0.9375$, $\epsilon(11) = 0.5343$. If interest lies in the two-factor interaction then the first design will be preferred to the second one. On the other hand, if efficient estimation of the main effects is considered more important, then the use of the second design will be advantageous.

Clearly, from (5.3.3), $0 \le \epsilon(x) \le 1$ for every x. Full information is retained on an interaction F^x if $\epsilon(x) = 1$. Suppose for any $i_1, \ldots, i_g$

$(1 \leq i_1 < \cdots < i_g \leq n, 1 \leq g \leq n)$, all the level combinations of the factors $F_{i_1}, \ldots, F_{i_g}$ occur equally often in the initial block(s) of a GC/n design (this implies that the same holds good for every other block of the design). Then such a design retains full information on all interactions involving $F_{i_1}, \ldots, F_{i_g}$. This follows immediately if one observes that by Theorem 4.5.1 (or its counterpart in the regular disconnected case), a GC/n design is always efficiency-consistent and that under the stated condition the incidence matrix of the subdesign corresponding to $F_{i_1}, \ldots, F_{i_g}$ is a multiple of that of a randomized (complete) block design. In fact, the converse is also true, i.e., if full information is retained in a GC/n design on all interactions involving $F_{i_1}, \ldots, F_{i_g}$, then in the initial block(s) of the design all the level combinations of these factors must occur equally often (see Mukerjee (1980) for details).

In particular, full information is retained on a main effect $F_i (1 \leq i \leq n)$ in a GC/n design if and only if the levels of the ith factor occur equally often in the initial blocks(s) of the design. For this, it is necessary that k should be a multiple of m_i. John (1973b) conjectured that if k is not a multiple of m_i, then one of the GC/n designs, where each level of F_i occurs either $[k/m_i]$ times or $[k/m_i]+1$ times in the initial block(s), will maximize the efficiency with respect to main effect F_i.

John (1981) considered the problem of balancing in GC/n designs. He observed that an interaction F^x is balanced in a GC/n design d, provided it is balanced in the corresponding subdesign d_x as defined in section 4.5. The proof is again straightforward since by Theorem 4.5.1, a GC/n design is efficiency-consistent. The finding is particularly helpful in constructing GC/n designs which are balanced with respect to one or more main effect(s).

<u>Example 5.3.3. (John, 1981).</u> Consider the construction of a $GC/2$ design with $m_1 = 5$, $m_2 = 4$, $r = k = 4$. Clearly the cyclic designs for 5 and 4 treatments with initial blocks (0,1,2,3) are both balanced. The $GC/2$ design with initial block {00,11,22,33} is, therefore, balanced for the main effects. It may, however, be seen that the design is not balanced for the two-factor interaction.

Lewis and Tuck (1985) considered paired comparison GC/n designs, i.e., GC/n designs with $k = 2$, which are balanced with respect to all the interactions. Clearly then, these designs are balanced with OFS. They supplied a table of such designs with $v(=\Pi m_i) \leq 36$, and illustrated their findings with a practical example. Recently, Gupta (1987c) developed an algorithm, based on the binary number association scheme, for the generation of balanced GC/n designs with $k = 2$.

Some further results on GC designs, not exactly in the context of factorial experiments, have been reported by Jarrett and Hall (1978).

5.4 Single replicate factorials in GC/n designs

Factorial designs with a single replicate are important in many practical situations where it is too expensive to have more than one replicate. John and Dean (1975) considered the construction of single replicate factorials in GC/n designs from a set of generating treatment combinations. The class of GC/n designs was seen to be flexible enough to provide a general method of confounding in symmetrical factorial experiments. The determination of the confounding scheme from an examination of the generating treatments was also demonstrated. Dean and John (1975) extended the results in John and Dean (1975) to asymmetrical factorial experiments. In this section, we shall consider single replicate factorials in GC/n designs mainly following Dean and John (1975). The results in John and Dean (1975) will follow by taking $m_1 = m_2 = \cdots = m_n$. It may be remarked that some of the results in this section were reviewed by Street (1986).

As usual, considering an n-factor setting with factors at $m_1, \ldots, m_n (\geq 2)$ levels, take a typical treatment combination $j = (j_1, \ldots, j_n)$. Let μ be the lowest common multiple of $m_1, \ldots, m_n$ and let

$$w = HCF(\mu, \mu j_1/m_1, \ldots, \mu j_n/m_n) \tag{5.4.1}$$

where HCF means highest common factor. Let the treatment combination obtained by multiplying each j_i by u mod m_i be denoted by uj. Since $\mu j/w \equiv 0$, a cyclic Abelian group S of size μ/w is given by the treatment combinations $0, j, 2j, \ldots, (\mu/w-1)j$. The treatment combination j is called a generator of S and the treatment combinations in S will constitute the initial block of a single replicate GC/n design with $k = \mu/w$.

More generally, consider f generators $b_1, \ldots, b_f$ with corresponding Abelian groups $S_1, \ldots, S_f$ where for each i, the intersection of S_i and the direct sum of the other groups is empty. Let w_i be defined as in (5.4.1) for b_i, let $q_i = \mu/w_i$ and $q = \prod_{i=1}^{f} q_i$. Then the group S formed by taking the direct sum of $S_1, \ldots, S_f$, with general element

$$u_1 b_1 + \cdots + u_f b_f \quad (0 \le u_i \le q_i - 1, 1 \le i \le f),$$

has q distinct elements. These treatment combinations constitute the initial block of a single replicate GC/n design with $k = q$. The remaining blocks can be easily obtained by taking a treatment combination not contained in a previous block and adding it in turn to each element of the initial block. Thus the remaining blocks are cosets of the subgroup S in the group of all treatment combinations.

Example 5.4.1 (Dean and John, 1975). Let $n = 3$, $m_1 = 3$, $m_2 = 4$, $m_3 = 6$. Then $\mu = 12$ and with $b_1 = 111$, $b_2 = 023$, one gets $w_1 = HCF(12,4,3,,2) = 1$, $w_2 = HCF(12,0,6,6) = 6$. Hence $q_1 = 12$, $q_2 = 2$. These generators, therefore, give rise to a single replicate $GC/3$ design for a $3 \times 4 \times 6$ factorial with k = 24.

Let C be the C-matrix of a single replicate GC/n design constructed as above. Note that $C = I - k^{-1}NN'$, where N is the incidence matrix of the design. Since $N'N = kI$, it follows that C is idempotent. Furthermore, it may be seen that C is of the form

$$C = I - k^{-1} \sum_{j_1=0}^{m_1-1} \cdots \sum_{j_n=0}^{m_n-1} \xi_{j_1 \cdots j_n} \left(\bigotimes_{i=1}^{n} R_{ij_i}^{*} \right), \tag{5.4.2}$$

where $\xi_{j_1 \cdots j_n} = 1$ if the treatment combination $(j_1, \ldots, j_n)$ occurs in the initial block, and zero otherwise. As in the preceding section, C has structure K so that by Theorem 4.3.2, the design is regular and has OFS. In order to determine the confounding scheme, let for each $x = (x_1, \ldots, x_n) \in \Omega$, V_x^* be the space of coefficient vectors of estimable contrasts belonging to the interaction F^x and

$$\bar{\alpha}(x) = k^{-1} \sum_{j_1=0}^{m_1-1} \cdots \sum_{j_n=0}^{m_n-1} \xi_{j_1 \ldots j_n} \left(\prod_{i=1}^{n} z_{j_i}^{x_i}\right), \tag{5.4.3}$$

where for $1 \leq i \leq n$,

$$\begin{aligned} z_{j_i}^{x_i} &= m_i - 1 && \text{if } j_i = 0,\ x_i = 1 \\ &= -1 && \text{if } j_i \neq 0,\ x_i = 1 \\ &= 1 && \text{if } x_i = 0. \end{aligned}$$

Then one obtains the following theorem which shows that for a single replicate GC/n design constructed as above, the number of degrees of freedom of the interaction F^x confounded with blocks is precisely the same as $\bar{\alpha}(x)$.

Theorem 5.4.1. For each $x \in \Omega$, rank $(V_x^*) = \alpha(x) - \bar{\alpha}(x)$, where $\alpha(x) = \Pi(m_i - 1)^{x_i}$.

Proof. Since C has structure K, rank $(V_x^*) = rank(P^x C)$, by Lemma 4.3.1. Now by (2.2.7), (2.2.8), $P^{x'} P^x C = M^x C$ and $P^x M^x C = P^x C$, so that $rank\ (P^x C) = rank(M^x C)$. Hence

$$rank(V_x^*) = rank(M^x C). \tag{5.4.4}$$

But C has structure K, so that $M^x C = CM^x$, by (4.2.3). Also by (2.2.1), (2.2.2), M^x is idempotent. Hence the fact that C is idempotent implies that $M^x C$ is idempotent. Consequently by (5.4.4),

$$rank(V_x^*) = tr(M^x C) = \alpha(x) - \bar{\alpha}(x),$$

after some simplification using (2.2.1), (2.2.2) and (5.4.2).

Following John and Dean (1975) and Dean and John (1975), it is possible to simplify the expression for $\bar{\alpha}(x)$ as given by (5.4.3). Note that $\bar{\alpha}(x)$ depends only on the treatment combinations in the initial block which in turn depend only on the generators of the block. It is, therefore, possible to obtain $\bar{\alpha}(x)$ directly from the generators. Consider the g-factor interaction involving the factors $F_{i_1}, \ldots, F_{i_g}$ $(1 \leq i_1 < \cdots < i_g \leq n, 1 \leq g \leq n)$, i.e., the interaction F^x where $x = (x_1, \ldots, x_n)$, with $x_i = 1$ if $i = i_1, \ldots, i_g$ and zero otherwise. Let

$$B = \begin{bmatrix} b_{1i_1} & b_{2i_1} & \cdots & b_{fi_1} \\ b_{1i_2} & b_{2i_2} & \cdots & b_{fi_2} \\ \cdot & \cdot & \cdots & \cdot \\ \cdot & \cdot & \cdots & \cdot \\ \cdot & \cdot & \cdots & \cdot \\ b_{1i_g} & b_{2i_g} & \cdots & b_{fi_g} \end{bmatrix},$$

where the treatment combinations $b_j = (b_{j1}, b_{j2}, \ldots, b_{jn})$ $(1 \le j \le f)$ generate the initial block. As seen earlier, the treatment combinations in the initial block are of the form

$$u_1 b_1 + \cdots + u_f b_f \quad (0 \le u_j \le q_j - 1,\ 1 \le j \le f).$$

Now the number of treatment combinations in the initial block with factors $F_{i_1}, \ldots, F_{i_g}$ all at zero levels will be the same as the number of solutions to the system of linear congruences

$$B\underline{u} \equiv \underline{0}, \tag{5.4.5}$$

where the jth row is reduced modulo $m_j (j = i_1, \ldots, i_g)$ and $\underline{u} = (u_1, \ldots, u_f)'$.

Let B^* be the matrix obtained by multiplying each element in the jth row of B by μ / m_j $(j = i_1, \ldots, i_g)$. Consider all the $t \times t$ submatrices contained in the $i_1^*th, \ldots, i_t^*th$ rows of B^* and let $d_{i_1^* \cdots i_t^*}$ be the highest common factor of the absolute values of their determinants $(t \le \min(g, f))$. Define

$$D_t = \begin{cases} 1, & \text{if } t = 0 \\ HCF(d_{i_1^* \cdots i_t^*} \mid \{i_1^*, \ldots, i_t^*\} \subset \{i_1, \ldots, i_g\}), & \text{if } 0 < t \le f \\ 0, & \text{if } t > f \end{cases}$$

Mathews (1892, p. 13) showed that the system of congruences (5.4.5) has the same number of solutions as the system $B^* \underline{u} = \underline{0}$, mod μ, i.e., each row reduced modulo μ. Following Smith (1861), Dean and John (1975) observed that the number of solutions is given by $\mu^{f-g} s_{i_1 \cdots i_g}$, where

$$s_{i_1 \cdots i_g} = \prod_{t=1}^{g} HCF(\mu, D_t/D_{t-1}), \text{ if } g \leq f \text{ and } D_g \neq 0$$
$$= \mu^{g-u} HCF(s_{i_1^* \cdots i_u^*} | \{i_1^*, \ldots, i_u^*\} \subset \{i_1, \ldots, i_g\}), \text{ if } g > u,$$

where u is such that $D_u \neq 0$ and $D_{u+1} = \cdots = D_g = 0$. Let $\overline{\alpha}(x)$ be denoted by $\overline{\alpha}_{i_1 \cdots i_g}$, where x has the $i_1 th, \ldots, i_g th$ digits unity and the remainder zero. Then it can be shown that for the g-factor interaction involving the factors $F_{i_1}, \ldots, F_{i_g}$, the number of degrees of freedom confounded with blocks is given by

$$\overline{\alpha}_{i_1 \cdots i_g} = (m_{i_1} m_{i_2} \cdots m_{i_g}/\mu^g) s_{i_1 \cdots i_g} - \sum_{u=1}^{g-1} (\overline{\alpha}_{i_1^* \cdots i_u^*} | \{i_1^*, \ldots, i_u^*\} \subset \{i_1, \ldots, i_g\}) - 1. \tag{5.4.6}$$

Since the practically useful designs are constructed from a small number of generators, it is worth noting that when $g > f$, (5.4.6) is particularly easy to calculate. The quantity $s_{i_1 \cdots i_g}$ will only involve the highest common factor of quantities previously calculated for f-factor interactions.

Example 5.4.2. Let $n = 3$, $m_1 = 2$, $m_2 = m_3 = 3$ and $f = 1$. The generator $b_1 = 111$ gives a design in 3 blocks of 6 treatment combinations each. Here $\mu = 6$, $d_1 = 3$, $d_2 = d_3 = 2$. Therefore, $s_1 = 3$, $s_2 = s_3 = 2$ and $\overline{\alpha}_1 = \overline{\alpha}_2 = \overline{\alpha}_3 = 0$. As for interaction F_1F_2, one has $D_0 = 1$, $D_1 = 1$, $D_2 = 0$, so that $s_{12} = \mu\ HCF(s_1, s_2) = 6$ and $\overline{\alpha}_{12} = 0$. Similarly, $s_{13} = 6$, $s_{23} = 12$, $s_{123} = 36$ and, therefore, $\overline{\alpha}_{13} = 0$, $\overline{\alpha}_{23} = 2$, $\overline{\alpha}_{123} = 0$. Thus in the design under consideration, 2 degrees of freedom from interaction F_2F_3 are confounded with blocks, while all contrasts belonging to the other main effects and interactions are estimable.

The relations (5.4.3), (5.4.6) are useful for computing $\overline{\alpha}(x)$, $x \in \Omega$, in single replicate GC/n designs. Several other procedures, including those following the lines of Bailey (1977), Collings (1984) and Voss and Dean (1988), have been discussed by Voss and Dean (1985) with examples.

Dean and John (1975) tabulated single replicate asymmetric factorial designs which leave all main effects free from block effects and which confound the highest order interaction possible for $n \leq 5$, $v = m_1 m_2 \cdots m_n \leq 56$, $k \leq 30$ and

$m_i \leq 7(1 \leq i \leq 5)$. For each such design, they also indicated the number of degrees of freedom from each interaction confounded with blocks. Lewis (1982) extended the table to $n \leq 7$, $v \leq 200$, $m_i \leq 7(1 \leq i \leq 7)$.

Dean (1978) considered the problem of partitioning the interactions of a single replicate factorial experiment in a GC/n design, where possible, into linear, quadratic, cubic and higher degree components each based on a single degree of freedom. Such partitioning is appropriate when the factors are quantitative. The results in Dean (1978) are essentially based on the fact that in a single replicate design, the best linear unbiased estimators (BLUE's) of mutually orthogonal estimable treatment contrasts are uncorrelated. This is shown in the following lemma.

Lemma 5.4.1. Let $\underline{h}_1' \underline{\tau}$, $\underline{h}_2' \underline{\tau}$ be any two mutually orthogonal estimable treatment contrasts in a single replicate factorial design. Then $Cov(\underline{h}_1' \hat{\underline{\tau}},\ \underline{h}_2' \hat{\underline{\tau}}) = 0$.

Proof. The C-matrix of a single replicate design is idempotent so that C itself is a g-inverse of C. Hence

$$Cov(\underline{h}_1' \hat{\underline{\tau}}, \underline{h}_2' \hat{\underline{\tau}}) = \sigma^2 \underline{h}_1' C \underline{h}_2. \tag{5.4.7}$$

where σ^2 is the error variance. For a single replicate design, it is easy to see that $CN = 0$. Since $\underline{h}_1' \underline{\tau}$ is estimable, it follows that $\underline{h}_1' N = 0$, i.e., $\underline{h}_1' C = \underline{h}_1'$. Hence by (5.4.7), $Cov(\underline{h}_1' \hat{\underline{\tau}}, \underline{h}_2' \hat{\underline{\tau}}) = 0$, since $\underline{h}_1' \underline{h}_2 = 0$, completing the proof of the lemma.

In a single replicate GC/n design for some $x \in \Omega$, suppose $rank(V_x^*)(= \alpha^*(x), say) > 0$ and let $\{\underline{h}_j, 1 \leq j \leq \alpha^*(x)\}$ represent an orthogonal basis for V_x^*. Then by Lemma 5.4.1,

$$SS \ \ due \ \ to \ \ F^x = \sum_{j=1}^{\alpha^*(x)} \{SS \ \ due \ \ to \ \ \underline{h}_j' \hat{\underline{\tau}}\},$$

where each SS in the right-hand member carries a single degree of freedom. If F^x is a main effect then one may select the $\{\underline{h}_j\}$ to represent its linear, quadratic, cubic or higher degree components provided such components are estimable. A table of orthogonal polynomials (Fisher and Yates, 1963) will be useful in this regard. Higher order interactions may also be handled in a similar manner. For further details, we refer to Dean (1978).

Generalizing an approach of Das (1964), Cotter (1974) suggested a method of constructing a single replicate m^n symmetric factorial design in m^{n-f} blocks, each of size m^f $(1 \leq f < n)$. It may be seen that the designs so constructed are GC/n designs. In fact, the method due to John and Dean (1975) or Dean and John (1975) covers that due to Cotter (1974) as a special case (see Voss and Dean (1987)) if one takes the generators as

$$b_1 = (1,0,\ldots,0,b_{1f+1},\ \ldots\ ,b_{1n}),$$
$$b_2 = (0,1,\ldots,0,b_{2f+1},\ \ldots\ ,b_{2n}),$$
$$\vdots$$
$$b_f = (0,0,\ldots,1,b_{ff+1},\ \ldots\ ,b_{fn}),$$

where the $b_{ii'}(1 \leq i \leq f, f+1 \leq i' \leq n)$ are to be suitably chosen depending on the desired pattern of confounding. As in the preceding section, full information will be retained on all interactions involving up to $g(\leq n)$ factors provided all level combinations of every g of the n factors occur equally often in the initial block, i.e., provided the $n \times m^f$ array obtained by writing the treatment combinations in the initial block as columns is an orthogonal array of strength g (Raghavarao, 1971). For several choices of $m, f, g,$ Cotter (1974) discussed the selection of the $b_{ii'}$ $(1 \leq i \leq f, f+1 \leq i' \leq n)$ so that the initial block may represent such an orthogonal array.

Lewis, Dean and Lewis (1983) presented single replicate designs involving two quantitative factors. The designs were selected from the individual replicates of resolvable $GC/2$ designs under the criterion of low loss of information on low degree components of the main effects and the two-factor interaction.

5.5. Further results

This section considers some combinatorial properties of GC/n designs. The results are due to Dean and Lewis (1980, 1986). From Section 5.3 recall that a GC/n design is obtained by taking full or fractional generalized cyclic (GC) sets individually or in combination and that a fractional GC set is one that has a fraction, $1/p$ say, of the v blocks replicated p times. In factorial experiments often the number of treatment combinations is large even for moderate numbers of factors and levels. In many cases, therefore, a full GC set of v blocks may be too large to be useful in practice. Theorem 5.5.1 below shows how to construct an initial block which will generate a GC/n design in a fractional GC set of v/p blocks, $p > 1$.

The v treatment combinations form an Abelian group, say T, under the operation of addition as defined in Section 5.3. Let A_1, A_2 be subsets of T such that all the elements in the set $A_{12} = \{a_1 + a_2 \colon a_1 \in A_1, a_2 \in A_2\}$ are distinct. Then we shall write $A_1 \oplus A_2$ for the set A_{12}.

Theorem 5.5.1 (Dean and Lewis, 1980). Cyclic development of an initial block generates a binary GC/n design with exactly v/p distinct blocks if and only if the initial block can be expressed as $T_1 \oplus T_2$ where

(i) T_1 is a subgroup of T whose order p is a common factor of k and v;
(ii) T_2 is a subset of T of size k/p;
(iii) There is no subgroup T_1^* of T of order greater than p for which the initial block may be expressed as $T_1^* \oplus T_2^*$, $T_2^* \subset T$.

When v is divisible by k a large number of GC/n designs are resolvable, i.e., the v/p blocks of the design can be grouped into k/p replicates of blocks (there being v/k blocks in each replicate) so that each replicate contains every treatment combination exactly once. The next two theorems give necessary and sufficient conditions for an initial block to generate a resolvable GC/n design. In the following for any subset A of T, D_A denotes the set of distinct differences $a_i - a_j$ $(a_i, a_j \in A)$.

Theorem 5.5.2 (Dean and Lewis, (1980). A binary GC/n design, with initial block $T_1 \oplus T_2$ as defined in Theorem 5.5.1, is resolvable if and only if k divides v and there exists a subset T_3 of T, of size v/k such that T may be represented as $T_1 \oplus T_2 \oplus T_3$.

Theorem 5.5.3 (Dean and Lewis, 1980). Let the initial block of a binary GC/n design be $T_1 \oplus T_2$ as defined in Theorem 5.5.1, and let T_3 be any subset of T of size v/k. The set T_3 satisfies the requirements of Theorem 5.5.2 if and only if D_{T_3} and $D_{T_2} \oplus T_1$ have only the zero element in common.

For the proofs of Theorems 5.5.1 - 5.5.3, we refer to Dean and Lewis (1980). Given the initial block of a GC/n design, these theorems enable the following questions to be answered with the minimum of search time: (i) How many distinct blocks will be generated? (ii) Will the design be resolvable? (iii) If the design is resolvable, what is the ith block of the jth replicate? The following procedure was suggested by Dean and Lewis (1980):

Step I. Find the largest subgroup T_1 of T such that the initial block can be expressed as a union of cosets of T_1. If T_1 has order p then the GC/n design will have v/p blocks.

Step II. Let the initial block be expressed as $T_1 \oplus T_2$. Compute the difference set D_{T_2} and the set $D_{T_2} \oplus T_1$. Determine the set $\overline{D}$ containing 0 and any element of T not in $D_{T_2} \oplus T_1$. Look for a set T_3 in $\overline{D}$ with v/k elements such that $D_{T_3} \subseteq \overline{D}$. Such a set may not exist in which case the GC/n design with the given initial block is not resolvable.

Step III. The ith block of the jth replicate of the design is obtained by adding $t_{3i} + t_{2j}$ to each treatment combination in the initial block, where $t_{3i} \in T_3$, $t_{2j} \in T_2$, $1 \le i \le v/k$, $1 \le j \le k/p$.

Example 5.5.1. (Dean and Lewis, 1980). Let $n = 2$, $m_1 = 2$, $m_2 = 6$, $k = 4$, and consider the initial block {00,02,11,13}. The subgroup {00,13} contained in the initial block does not allow the initial block to be expressed in the required form. The largest allowable subgroup is $T_1 = \{00\}$ with $T_2 = \{00,02,11,13\}$. Hence $p = 1$ and there are 12 distinct blocks in the design. Since $D_{T_2} = \{00,02,11,13,04,15\} = D_{T_2} \oplus T_1$, there does not exist a set T_3 in $\overline{D}$ of size

$v/k = 3$ satisfying $D_{T_3} \cap (D_{T_2} \oplus T_1) = \{00\}$. Hence the GC/2 design is not resolvable.

With reference to Step II, Dean and Lewis (1980) suggested a method for finding the set T_3 if it exists. They also considered the problem of choosing the initial block so as to obtain a resolvable GC/n design when such a design exists. We refer to their original paper for the details in this regard. Certain aspects relating to efficiencies in GC/n designs were discussed by Dean and Lewis (1980); these will be considered later in Chapter 8. For a discussion on a related applied problem, reference is made to Lewis and Dean (1980).

Dean and Lewis (1986) considered the problem of connectedness in GC/n designs and obtained the following result.

Theorem 5.5.4 (Dean and Lewis, 1986). A binary GC/n design, with initial block $T_1 \oplus T_2$ as defined in Theorem 5.5.1, is connected if and only if $<T_2> + T_1 = T$, where $<T_2> + T_1$ is the set of distinct elements $t_2 + t_1$, $t_2 \in <T_2>$, $t_1 \in T_1$, and $<T_2>$ is the group generated by T_2.

Proof. Sufficiency: Let $<T_2> + T_1 = T$. Then every element of T can be expressed as a linear combination of the treatment combinations in $T_1 \oplus T_2$. Let t^*, t^{**} be any two distinct elements in T, and let $t^* - t^{**} = \sum_{i=1}^{k} \rho_i \varsigma_i$, where $\rho_1, \ldots, \rho_k$ are integers and $\varsigma_1, \ldots, \varsigma_k$ are the treatment combinations in $T_1 \oplus T_2$. The blocks of a GC/n design are obtained by adding each element of T to $T_1 \oplus T_2$. Hence one may construct a chain of treatment combinations, namely,

$$t^{**}, t^{**} + \varsigma_1, t^{**} + 2\varsigma_1, \ldots, t^{**} + \rho_1\varsigma_1, t^{**} + \rho_1\varsigma_1 + \varsigma_2, \ldots, t^{**} + \sum_{i=1}^{k} \rho_i \varsigma_i = t^*, \tag{5.5.1}$$

such that every two consecutive members of the chain occur together in some block. It follows that the design is connected.

Necessity: Let $<T_2> + T_1$ be a proper subset and hence a proper subgroup of T.

Then there exists at least one other coset, say T^*, of $<T_2> + T_1$ in T. Since the intersection of $<T_2> + T_1$ and T^* is empty, there can be no chain like (5.5.1) linking an element of $<T_2> + T_1$ to an element in T^*.

For a disconnected GC/n design, the confounded contrasts may be obtained by computing the annihilators of $<T_2> + T_1$ (see Dean and Lewis (1984) for more details).

Example 5.5.2. (Dean and Lewis, 1986). With $n = 2$, $m_1 = m_2 = 6$, suppose a connected design is required in 18 blocks each of size 8. A $GC/2$ design in 18 blocks can be obtained by selecting an initial block $T_1 \oplus T_2$ such that T_1 is of order 2 and T_2 is of order 4. Let $T_1 = \{00,33\}$ and consider the following three possibilities for T_2:

$$T_{21} = \{00,12,21,44\}, \quad T_{22} = \{00,15,44,53\}, \quad T_{23} = \{00,12,21,45\}$$

Since the cardinalities of $<T_{21}> + T_1$, $<T_{22}> + T_1$, $<T_{23}> + T_1$ equal 36, 18, 12 respectively, it follows that only T_{21} leads to a connected design.

We conclude this section with the remark that Theorems 5.5.1 - 5.5.4 can be easily extended to non-binary designs as indicated by Dean and Lewis (1980, 1986).

5.6. Row-column designs

John and Lewis (1983) introduced generalized cyclic row-column designs for factorial experiments. They considered an arrangement of the $v = \Pi m_j$ treatment combinations in g_1 rows and g_2 columns such that (a) a single treatment combination is applied to each of the $g_1 g_2$ cells of the design, (b) the treatment combinations are replicated the same number of times (hence $g_1 g_2$ must be a multiple of v), (c) the rows form a GC/n set, and (d) the columns from a GC/n set.

Of course, it is necessary to amalgamate the row and column components to give a row-column design. Recall that the v treatment combinations form an Abelian group, T, under the operation of addition. With reference to Theorem 5.5.1, suppose the initial row consists of a subgroup T_1 of order p together with a further $(g_2/p)-1$ of its cosets. Since a GC/n set of v/p blocks is generated from this row by the addition of one treatment combination from each of the cosets of T_1, the initial column must consist of g_1p/v elements from each of the cosets of T_1. By a similar argument, the initial column must consist of a subgroup T_1' of order p' together with a further $(g_1/p')-1$ of its cosets and that the initial row must consist of g_2p'/v elements from each of the cosets of T_1'. Note that p and p' must be common factors of v and g_2 and of v and g_1 respectively. Also v must be a divisor of g_1p and g_2p'. Under these conditions, the row-column design contains $r = g_1g_2/v$ replicates of each treatment combination.

Example 5.6.1. (John and Lewis, 1983). Let $n = 3$, $m_1 = 2$, $m_2 = 3$, $m_3 = 5$, $g_1 = 6$, $g_2 = 10$. Suppose that the row component is obtained from an initial block consisting of a subgroup $T_1 = \{000,001,002,003,004\}$ and the coset $\{110,111,112,113,114\}$ and that the column component is obtained from an initial block consisting of a subgroup $T_1' = \{000,010,020\}$ together with the coset $\{101,111,121\}$. Here $p = 5$, $p' = 3$, $g_1p/v = 1$, $g_2p'/v = 1$. Thus the initial column (row) consists of 1 element from each coset of $T_1(T_1')$. The row and column components can, therefore, be amalgamated to form the following row-column design where each treatment combination is replicated twice.

000	001	002	003	004	110	111	112	113	114
010	011	012	013	014	120	121	122	123	124
020	021	022	023	024	100	101	102	103	104
101	102	103	104	100	011	012	013	014	010
111	112	113	114	110	021	022	023	024	020
121	122	123	124	120	001	002	003	004	000

The intrablock matrix of a $g_1 \times g_2$ row column design is given by

$$C = rI - g_2^{-1}N_1N_1' - g_1^{-1}N_2N_2' + (r^2/g_1g_2)J,$$

where $r(=g_1g_2/v)$ is the common replication number, J is a $v \times v$ matrix of unit elements, and N_1 and N_2 are treatment-versus-row and treatment-versus-column incidence matrices respectively. Let $C_1 = rI - g_2^{-1}N_1N_1'$ and $C_2 = rI - g_1^{-1}N_2N_2'$ be the C-matrices of the row and column component designs respectively. Then

$$C = C_1 + C_2 - rI + (r^2/g_1g_2)J. \tag{5.6.1}$$

Under the present method of construction, both the row component and the column component represent GC/n designs. Consequently, both N_1N_1' and N_2N_2' are of the form (5.3.1) and hence both C_1 and C_2 have structure K. It follows from (5.6.1) that C has structure K so that a generalized cyclic row-column design has OFS (and is regular, if disconnected). The following lemma is helpful in the computation of the interaction efficiencies.

Lemma 5.6.1. For each $x \in \Omega$, $P^x C_1 P^{x'}$ and $P^x C_2 P^{x'}$ commute with each other.

Proof. Since both N_1N_1' and N_2N_2' are of the form (5.3.1) the matrices C_1 and C_2 commute with each other. Also both C_1 and C_2 have structure K. Hence by (2.2.7) and Theorem 4.2.4

$$\begin{aligned} P^x C_1 P^{x'} P^x C_2 P^{x'} &= P^x C_1 M^x C_2 P^{x'} = P^x M^x C_1 C_2 P^{x'} = P^x M^x C_2 C_1 P^{x'} \\ &= P^x C_2 M^x C_1 P^{x'} = P^x C_2 P^{x'} P^x C_1 P^{x'}. \end{aligned}$$

By Lemma 5.6.1, for each $x \in \Omega$, $P^x C_1 P^{x'}$ and $P^x C_2 P^{x'}$ have $\alpha(x)$ $(= \Pi(m_i - 1)^{x_i})$ common orthonormal eigenvectors (Rao, 1973a, p. 41). Hence by (5.6.1.), the eigenvalues of $P^x C P^{x'}$ are given by

$$\lambda_j^x = \lambda_{1j}^x + \lambda_{2j}^x - r, \tag{5.6.2}$$

where λ_{1j}^x and λ_{2j}^x are eigenvalues of $P^x C_1 P^{x'}$ and $P^x C_2 P^{x'}$ respectively, corresponding to the same eigenvector $(1 \le j \le \alpha(x))$. One may employ relations analogous to (5.3.5) to compute λ_{1j}^x, λ_{2j}^x, and then (5.3.3), (5.6.2) may be used for the computation of the interaction efficiencies. It follows from (5.6.2) that in order to construct a useful row-column design, the row and column components should be chosen to have high efficiencies with respect to the interactions of interest. In particular, as in Section 5.3, the row-column design retains full information on all interactions involving the factors $F_{i_1}, \ldots, F_{i_g}$ $(1 \le i_1 < \cdots < i_g \le n, 1 \le g \le n)$ provided all the level combinations of $F_{i_1}, \ldots, F_{i_g}$ occur equally often in the initial blocks of the row and column components. For further discussion on the choice of row and column components with examples, we refer to John and Lewis (1983).

Consider now a single replicate generalized cyclic row-column design. For each $x \in \Omega$, let $\overline{\alpha}_0(x)$, $\overline{\alpha}_1(x)$, $\overline{\alpha}_2(x)$ denote the numbers of degrees of freedom belonging to the interaction F^x confounded with the row-column design, the row component design and the column component design respectively. Note that $\overline{\alpha}_1(x)$, $\overline{\alpha}_2(x)$ may be computed using (5.4.6). The following theorem gives a simple rule for obtaining $\overline{\alpha}_0(x)$ from $\overline{\alpha}_1(x)$ and $\overline{\alpha}_2(x)$.

Theorem 5.6.1 (John and Lewis, 1983). For each $x \in \Omega$, $\overline{\alpha}_0(x) = \overline{\alpha}_1(x) + \overline{\alpha}_2(x)$.

Proof. As seen above, each of C, C_1, C_2 has structure K. Also for a single replicate design, it is easy to check that each of them is idempotent. Hence as in the proof of Theorem 5.4.1,

$$\alpha(x) - \overline{\alpha}_0(x) = tr(M^x C), \;\; \alpha(x) - \overline{\alpha}_i(x) = tr(M^x C_i) \;\; (i{=}1{,}2). \qquad (5.6.3)$$

By (2.2.1), (2.2.2), $tr(M^x) = \alpha(x)$. Hence the result follows from (5.6.1.) and (5.6.3).

The ideas in John and Lewis (1983) were extended further by Lewis (1986) who considered composite generalized cyclic row-column designs. These are designs for which both the row component and the column component are unions of one or more GC/n sets of blocks. Using arguments similar to those in John and Lewis (1983), as indicated earlier in this section, Lewis (1986) discussed the problem of amalgamation of the two component designs.

Example 5.6.2. (Lewis 1986). A composite generalized cyclic row-column design with $n = 2$, $m_1 = 2$, $m_2 = 4$, $g_1 = 4$, $g_2 = 6$ is shown below. Here the row component design is a single $GC/2$ set of 4 blocks with the initial block consisting of the subgroup {00,10} and the cosets {01, 11}, {02, 12}. The column component design is composed of a $GC/2$ set of 2 blocks (columns 1 and 2) together with a $GC/2$ set of 4 blocks. The initial block of the first $GC/2$ set of the column component is given by the subgroup {00, 11, 02, 13} and the initial block of the second $GC/2$ set of the column component consists of the subgroup {00, 12} together with the coset {01, 13}.

00	10	01	11	02	12
11	01	12	02	13	03
02	12	13	03	10	00
13	03	00	10	01	11

In general, amalgamation of the row and column component designs may be achieved using the algorithm of Jones (1980). As Lewis (1986) indicates, for small layouts, amalgamation by hand is not too tedious although some trial and error is involved. In particular, if all the GC/n sets in the row component are identical and all the GC/n sets in the column component are identical, then generalized cyclic row-column designs as in John and Lewis (1983) are obtained.

Analogously to (5.6.1), the C-matrix of a composite generalized cyclic row-column design can be expressed as a linear combination of C_1, C_2, I and J, where C_1, C_2 are the C-matrices of the row and column component designs respectively. Since C_1 and C_2 have structure K, C will also have structure K and hence a composite generalized cyclic row-column design has OFS (and is regular, if disconnected). As before, Lemma 5.6.1 holds and the interaction efficiencies in a composite row-column design may be obtained from consideration of the row and column component designs.

5.7. Designs with partial orthogonal factorial structure.

Chauhan and Dean (1986) discussed the construction of an $m_1' \times m_2 \times \cdots \times m_n'$ factorial design with partial orthogonal factorial structure (POFS) with respect to several interactions from an $m_1 \times m_2 \times \cdots \times m_n$ factorial design with OFS, where $\Pi m_i' = \Pi m_i$, by factorizing and/or combining factor levels. As an illustration, they considered a 4×4 factorial in a $GC/2$ design, and mapped the levels of the second factor to the levels of two new factors each at two levels. The resulting $4 \times 2 \times 2$ factorial design was seen to have POFS with respect to most of the interactions.

In general, consider an $m_1 \times m_2 \times \cdots \times m_n$ factorial in a GC/n design and suppose that the levels of the nth factor are mapped to the levels of two new factors at m_n' and m_{n+1} levels respectively, where $m_n = m_n' m_{n+1}$.

Theorem 5.7.1 (Gupta, 1987b). For the $m_1 \times m_2 \times \cdots \times m_{n-1} \times m_n' \times m_{n+1}$ factorial design constructed as above:

(i) POFS holds with respect to all interactions which do not involve the nth factor.

(ii) For every interaction involving some or all of the first (n-1) factors, the efficiency in the derived design is the same as that in the original GC/n design.

(iii) POFS holds with respect to main effect F_n and interactin $F_n F_{n+1}$ if the subdesign (in the sense of Section 4.5) corresponding to the last two factors has OFS.

For a proof of Theorem 5.7.1, we refer to Gupta (1987b) where some further generalizations - in terms of mapping the levels of more than one factor - have also been considered.

CHAPTER 6
CONSTRUCTIONS II: DESIGNS BASED ON KRONECKER TYPE PRODUCTS

6.1. Introduction

As discussed in the last chapter, cyclic and generalized cyclic designs provide a wide class of designs for factorial experiments with OFS. Since the characterization for OFS as in Theorem 4.2.1 intrinsically involves Kronecker products, it is also appropriate to consider methods of construction based on Kronecker or Kronecker-type products of varietal designs. Such methods also lead to a wide variety of designs with OFS and, if appropriately used, are capable of ensuring high efficiencies with respect to the interactions of interest. The work presented here is mainly due to Mukerjee (1981a, 1984, 1986), Gupta (1983b, 1985, 1986a), Mukerjee and Sen (1987) and Mukerjee and Bose (1988b). It may be remarked that Kronecker designs, not exactly in the context of factorial experiments, were earlier considered by Vartak (1955).

We shall consider the construction of an $m_1 \times m_2 \times \cdots \times m_n$ factorial design D by taking a Kronecker or Kronecker-type product of varietal designs $D_1, \ldots, D_n$ involving $m_1, \ldots, m_n$ varieties respectively. It will be seen that by choosing $D_1, \ldots, D_n$ appropriately it is possible to ensure high interaction efficiencies in D. To that effect, the following notions will be helpful.

To begin with, let $D_1, \ldots, D_n$ be equireplicate with common replication numbers $r_1, r_2, \ldots, r_n$ respectively. For $1 \leq i \leq n$, let C_i be the C-matrix of D_i and $\lambda_{io} = 0$, λ_{it} $(1 \leq t \leq m_i - 1)$ be the eigenvalues of C_i. Then following Kiefer (1975), the Φ_p-efficiency of D_i is given by, say,

$$\left.\begin{aligned} E_{ip} &= \{ \prod_{t=1}^{m_i-1} (\lambda_{it}/r_i)\}^{(1/(m_i-1))} && \text{when } p = 0 \\ &= \{(m_i-1)^{-1} \sum_{t=1}^{m_i-1} (\lambda_{it}/r_i)^{-p}\}^{-1/p} && \text{when } 0 < p < \infty \\ &= \min_{1\le t\le m_i-1} (\lambda_{it}/r_i) && \text{when } p = \infty \end{aligned}\right\} \quad (6.1.1)$$

provided that $\lambda_{it}(1\le t\le m_i-1)$ are all positive, $1 \le i \le n$. If $\lambda_{it}(1\le t\le m_i-1)$ are not all positive, in which case D_i is disconnected, we define $E_{ip} = 0(0\le p\le\infty)$.

Clearly, if $p = 0,1,\infty$, then Φ_p-efficiency reduces to the standard $D-$, $A-$, $E-$efficiencies respectively. The definition of efficiency as considered in the preceding chapter (see (5.3.3)) corresponds to $p = 1$. In particular, if D_i be variance balanced then $\lambda_{it} = \lambda_i$, say $(1\le t\le m_i-1)$, and $E_{ip} = \lambda_i/r_i$, which is termed simply the efficiency of the design.

Turning to the n-factor design D, recall that by Lemma 2.2.3, for each $x \in \Omega$, $P^x\underline{\tau}$ represents a complete set of orthonormal contrasts belonging to the interaction F^x. Let A_x be the coefficient matrix of the reduced normal equations for estimating $P^x\underline{\tau}$ in D, i.e., A_x is the information matrix for $P^x\underline{\tau}$ in D. Let $\lambda_t^x(1\le t\le\alpha(x))$, where $\alpha(x) = \Pi(m_i-1)^{x_i}$, be the eigenvalues of A_x. Then the Φ_p-efficiency, say E_p^x, of D with respect to the interaction F^x may be defined analogously to (6.1.1) replacing m_i-1, r_i, $\{\lambda_{it}, 1\le t\le m_i-1\}$ there by $\alpha(x)$, r and $\{\lambda_t^x, 1\le t\le\alpha(x)\}$ respectively, where r is the common replication number in D. In the trivial case where the $\{\lambda_t^x\}$ are not all positive, we define $E_p^x = 0(0\le p\le\infty)$. If the effect F^x is balanced in D, i.e., if all the $\{\lambda_t^x\}$ are equal then E_p^x is the same for all p, the common value being the efficiency of F^x in D.

6.2. Designs through ordinary Kronecker product

As in Section 6.1, for $1 \le i \le n$, let D_i be a varietal block design in m_i treatments with a common replication number r_i, a constant block size k_i and an incidence matrix N_i. Then

$$N^{(1)} = \bigotimes_{i=1}^{n} N_i$$

is the incidence matrix of a Kronecker block design in $v = \Pi m_i$ treatments with a common replication number Πr_i and a constant block size Πk_i. Interpreting the $v = \Pi m_i$ treatments in the Kronecker design in the lexicographic order, it represents an $m_1 \times m_2 \times \cdots \times m_n$ factorial design. Since

$$N^{(1)}N^{(1)'} = \bigotimes_{i=1}^{n} (N_i N_i'),$$

it follows from Theorem 4.2.2, 4.3.2 that the n-factor design given by $N^{(1)}$ has OFS. This is because for each i, D_i is equireplicate and has a constant block size so that $N_i N_i'$ is a proper matrix. Moreover, the following result holds on the interaction efficiencies, E_p^x, in the Kronecker design:

Theorem 6.2.1. For every $x \in \Omega$ and every $p\,(0 \leq p \leq \infty)$,

$$E_p^x \geq \max_{1 \leq i \leq n} \{x_i E_{ip}\},$$

and when F^x represents the ith main effect, $E_p^x = E_{ip}$.

A more general version of the above theorem has been proved in the next section. Thus using the ordinary Kronecker product, it is possible to ensure high interaction efficiencies by choosing the initial varietal designs $D_1, \ldots, D_n$ in a suitable manner. A difficulty with this method is that the block size and the number of replicates in design $N^{(1)}$ may become too large. For example, in an n-factor design based on ordinary Kronecker product the block size must be at least 2^n in order that all the main effect contrasts are estimable.

As an alternative approach one may, therefore, attempt to achieve a control only over the lower order interactions and successfully reduce the block size and/or the number of replicates. To that effect, some modifications of the ordinary Kronecker product may be employed which are discussed in the next sections.

6.3. Componentwise Kronecker product of order g

With $N_1, N_2, \ldots, N_n$ as before, let for $1 \leq i \leq n$,

$$N_i = N_{io} + N_{i1} + \cdots + N_{is_i-1}, \tag{6.3.1}$$

where u_i is a positive integer and the elements of N_{ih_i} $(0 \leq h_i \leq u_i - 1)$ are non-negative integers.

Definition 6.3.1. The componentwise Kronecker product of order $g (\leq n)$ of $N_1, N_2, \ldots, N_n$ with respect to the decomposition (6.3.1) is given by

$$N^{(2)} = \sum_{h_1, \ldots, h_n \in T} (\bigotimes_{i=1}^{n} N_{ih_i}). \tag{6.3.2}$$

the sum being taken over only a subset T of the Πu_i possible combinations $(h_1, h_2, \ldots, h_n)$ such that the combinations included in T, written as columns, form an orthogonal array (with possibly variable symbols) with n rows, $u_1, \ldots, u_n$ symbols and strength g (cf. Rao (1973b)).

The special case $g = n$ corresponds to the ordinary Kronecker product as considered in the last section. Mukerjee (1981a) considered the case of $g = 1$. Gupta (1987a) showed that if D_i be connected with $r_i = k_i = 2$ $(1 \leq i \leq n)$ then the design $N^{(2)}$ obtained using componenentwise Kronecker product of order 1 is connected if and only if $m_1, m_2, \ldots, m_n$ are relatively prime.

Theorem 6.3.1 If for each i, $h_i (0 \leq h_i \leq u_i - 1, 1 \leq i \leq n)$, the design represented by N_{ih_i} be equireplicate with replication number $u_i^{-1} r_i$ and has constant block size $u_i^{-1} k_i$, then the design $N^{(2)}$ obtained using the method of componentwise Kronecker product of order g has OFS with $E_p^x \geq \max_{1 \leq i \leq n} \{x_i E_{ip}\}$, $0 \leq p \leq \infty$, for every $x \in \Omega_g^*$, as before Ω_g^* being a subset of Ω containing vectors with at most g 1's. Furthermore, when F^x represents the ith main effect, $E_p^x = E_{ip}$.

Proof. Under the conditions of theorem for $0 \leq h_i,\ q_i \leq u_i - 1,\ 1 \leq i \leq n$, $N_{ih_i} N_{iq_i}'$ has all row and column sums equal. Further, the design $N^{(2)}$ is equireplicate with replication number $s \Pi u_i^{-1} r_i$ $(= r^{(2)}$, say) and has constant block size $s \Pi u_i^{-1} k_i$, s being the cardinality of T. Since by Definition 6.3.1,

$$N^{(2)} N^{(2)\prime} = \sum_{h, q \in T} (\bigotimes_{i=1}^{n} N_{ih_i} N_{iq_i}'), \tag{6.3.3}$$

where $h = (h_1, h_2, \ldots, h_n)$, $q = (q_1, q_2, \ldots, q_n)$, it is clear by Theorem 4.2.2 that the design represented by $N^{(2)}$ has OFS. The intrablock matrix C of the design $N^{(2)}$ is, say,

$$C = r^{(2)}(\bigotimes_{i=1}^{n} I_i) - s^{-1}(\prod_{i=1}^{n} u_i k_i^{-1})N^{(2)}N^{(2)'}, \tag{6.3.4}$$

I_i being the identity matrix of order m_i. To prove the rest of the theorem take any interaction F^x involving $f(\leq g)$ factors. Let without loss of generality, $x = \theta = (\theta_1, \theta_2, \ldots, \theta_n)$ with $\theta_i = 1$ $(1 \leq i \leq f)$, $= 0 (f+1 \leq i \leq n)$. Then by (2.2.5), (2.2.6), (6.3.3) and the assumptions regarding N_{ih_i},

$$P^\theta N^{(2)}N^{(2)'}P^{\theta'} = (\prod_{i=f+1}^{n} u_i^{-2}k_i r_i)(\bigotimes_{i=1}^{f} P_i)\{\sum_{h,q\in T}(\bigotimes_{i=1}^{f} N_{ih_i}N_{iq_i}')\}(\bigotimes_{i=1}^{f} P_i'). \tag{6.3.5}$$

Since $f \leq g$, (6.3.1) and the fact that the s members of T form an orthogonal array of strength g yield,

$$\sum_{h,q\in T}(\bigotimes_{i=1}^{f} N_{ih_i}N_{iq_i}') = \{s/(\prod_{i=1}^{f} u_i)\}^2(\bigotimes_{i=1}^{f} N_i N_i').$$

Hence if one defines the C-matrix of D_i as $C_i = r_i I_i - k_i^{-1}N_i N_i'$, employs (6.3.4), (6.3.5) and recalls that for a design with OFS, $A_x = P^x C P^{x'}$ for each $x \in \Omega$, then one obtains after some simplification,

$$A_\theta = P^\theta C^{(2)} P^{\theta'} = r^{(2)}[I - \bigotimes_{i=1}^{f}\{P_i(I_i - r_i^{-1}C_i)P_i'\}],$$

where I is the identity matrix of order $\alpha(\theta)$. Denote by $\lambda_{io} = 0$, $\lambda_{i1}, \ldots, \lambda_{i,m_i-1}$ the eigenvalues of C_i. Then it follows that the eigenvalues of A_θ are say,

$$\lambda^\theta_{t_1 t_2 \cdots t_f} = r^{(2)}[1 - \prod_{i=1}^{f}\{1 - (\lambda_{it_i}/r_i)\}],\ 1 \leq t_i \leq m_i - 1,\ 1 \leq i \leq f \tag{6.3.6}$$

Clearly for each $t_1, t_2, \ldots, t_f$,

$$\lambda^\theta_{t_1 t_2 \cdots t_f}/r^{(2)} \geq \max_{i \leq i \leq f}(\lambda_{it_i}/r_i) \tag{6.3.7}$$

From (6.3.6), (6.3.7), (6.1.1) and its analogue for factorial designs, it is clear that for the effect F^θ under consideration $E_p^\theta \geq \max_{1\leq i \leq f} E_{ip}$ $(0 \leq p \leq \infty)$. On similar lines it follows that when θ represents the ith main effect, $E_p^\theta = E_{ip}$. Hence the theorem.

Remark. Relations like (6.3.6) may be used for exactly determining E_p^x, $x \in \Omega_g^*$. Interestingly, as Example 6.3.1 below suggests, the exact values of E_p^x are usually much greater than the lower bound as given by Theorem 6.3.1. If the designs $D_1, D_2, \ldots, D_f$ are balanced then $\lambda_{i1} = \cdots = \lambda_{i,m_i-1} = \lambda_i$, say, and by (6.3.6), the effect F^{θ} is balanced with efficiency $1 - \prod_{i=1}^{f}(1-E_i)$, where $E_i\ (= \lambda_i/r_i)$ is the efficiency of D_i $(1 \le i \le f)$. This holds in general for any F^x, $x \in \Omega_g^*$, provided the relevant component designs are balanced.

A simple procedure for obtaining the matrices N_{ih_i} $(1 \le i \le n,\ 0 \le h_i \le u_i - 1)$ such that the conditions of Theorem 6.3.1 are satisfied is as follows. For $1 \le i \le n$, let the varieties in D_i be represented by the symbols 0, 1, ..., $m_i - 1$ and let z_i be an array obtained by writing the blocks of D_i as columns. Suppose for $1 \le i \le n$, (a) k_i, r_i are integral multiples of u_i and (b) partitioning z_i' into u_i subarrays each with $u_i^{-1}k_i$ columns as,

$$z_i' = (z_{io}', \ldots, z_{i,u_i-1}')$$

each symbol 0, 1, ..., $m_i - 1$ is repeated $u_i^{-1}r_i$ times in z_{ih_i} $(0 \le h_i \le u_i - 1)$. If for each i, h_i, N_{ih_i} be the incidence matrix of the design given by the columns of z_{ih_i}, then it is easily seen that the matrices N_{ih_i} satisfy the conditions of Theorem 6.3.1. In particular, if for some i, $k_i = u_i$, and D_i be a positionally balanced design (in the sense that in each row of z_i the symbols 0, 1, ..., $m_i - 1$ occur an equal number of times) then (a) holds trivially and (b) can be satisfied easily by taking z_{ih_i} as the $h_i th$ row of z_i $(0 \le h_i \le u_i - 1)$. Since it is easy to find a design D_i satisfying (a), (b), the method based on componentwise Kronecker product of order g becomes widely applicable. By suitably choosing $D_i (1 \le i \le n)$ it is often possible to ensure connectedness of the derived factorial design $N^{(2)}$. The following example illustrates the method.

Example 6.3.1. To construct a $3 \times 4 \times 5$ design, let D_1, D_2, D_3 be such that the arrays z_i, obtained by writing the blocks of D_i as columns, are as follows:

$$z_1 = \begin{matrix} 0\,1\,2 \\ 1\,2\,0 \end{matrix}, \quad z_2 = \begin{matrix} 0\,1\,2\,3 \\ 1\,2\,3\,0 \end{matrix}, \quad z_3 = \begin{matrix} 0\,1\,2\,3\,4\,0\,1\,2\,3\,4 \\ 1\,2\,3\,4\,0\,2\,3\,4\,0\,1 \end{matrix}$$

Then $r_1 = r_2 = 2$, $r_3 = 4$, $k_1 = k_2 = k_3 = 2$. Note that with blocks written as columns all the treatments occur equally often in each row. Hence if for each i, z_i be partitioned as $z_i = [z_{io}', z_{i1}']'$, where z_{ih_i} is a row-vector, and N_{ih_i} be the incidence matrix of a varietal design with blocks given by the columns of z_{ih_i} $(h_i=0,1)$ then the matrices N_{ih_i} satisfy the conditions of Theorem 6.3.1 with $u_1 = u_2 = u_3 = 2$. According to Definition 6.3.1, now $N^{(2)}$ may be formed by taking $T = \{(0,0,0),(0,1,1),(1,0,1),(1,1,0)\}$. Since T represents an orthogonal array of strength 2, the resulting $3 \times 4 \times 5$ factorial design has OFS and Theorem 6.3.1 holds with $g = 2$, i.e. for $0 \le p \le \infty$, Φ_p-efficiency of a main effect equals the Φ_p-efficiency of the relevant component design, and Φ_p-efficiency of a two-factor interaction is at least as large as the Φ_p-efficiencies of the relevant component designs. As D_1, D_3 are balanced with efficiencies 0.75 and 0.625, the remark following Theorem 6.3.1 shows that the effects F_1, F_3, F_1F_3 are balanced with respective efficiences 0.75, 0.625, 0.9062. The Φ_p-efficiencies of F_2, F_1F_2, F_2F_3 may be obtained by (6.3.6). In particular with $p = 1$, the A-efficiencies of these effects are 0.6, 0.9130, 0.8667 respectively. The design involves 8 replications and block size 4, while the corresponding usual Kronecker product requires 16 replications and block size 8.

Alternatively, taking D_1, D_2, D_3 such that

$$z_1 = \begin{matrix} 0 \\ 1 \\ 2 \end{matrix}, \quad z_2 = \begin{matrix} 0\,1\,2\,3 \\ 1\,2\,3\,0 \\ 2\,3\,0\,1 \end{matrix}, \quad z_3 = \begin{matrix} 0\,1\,2\,3\,4 \\ 1\,2\,3\,4\,0 \\ 2\,3\,4\,0\,1 \end{matrix},$$

and with $u_1 = u_2 = u_3 = 3$,
$T = \{(0,0,0),(0,1,1),(0,2,2,),(1,0,1),(1,1,2),(1,2,0),(2,0,2),(2,1,0),(2,2,1)\}$ one may proceed exactly as before to construct a $3 \times 4 \times 5$ factorial design in 3 replicates and block size 9. Since T is again an orthogonal array of strength 2, the factorial design obtained has OFS and Theorem 6.3.1 holds with $g = 2$. In this design the effects F_1, F_2, F_1F_2 are balanced with respective efficiencies 1, 0.8889, 1, while the A-efficiencies of the effects F_3, F_1F_3, F_2F_3 are 0.8148, 1, 0.9813 respectively. The design, like the earlier one is connected.

The second design in this example requires a smaller number of replications but a larger block size than the first one. In fact, many other $3 \times 4 \times 5$ designs with a fairly wide range of parameter-values and efficiency-levels can be obtained by the above method by choosing D_1, D_2, D_3 suitably. In a practical situation, a choice amongst the available designs depends on the particular context.

6.4. Khatri-Rao product of order g

Khatri and rao (1968) considered, in a different context, another modification of the ordinary Kronecker product. A generalized version of the Khatri-Rao product is considered below for the construction of factorial designs.

For $1 \leq i \leq n$, let D_i and N_i be as in the previous section, and partition N_i as

$$N_i = \bigcup_{h_i=0}^{u_i-1} N_{ih_i} \tag{6.4.1}$$

where N_{ih_i} is of order $m_i \times u_i^{-1} b_i$ $(0 \leq h_i \leq u_i - 1)$, b_i is the number of columns in N_i, u_i is a positive integer and for matrices $M_1, M_2, \ldots, M_a$ having the same number of rows,

$$\bigcup_{l=1}^{a} M_l = (M_1, M_2, \ldots, M_a).$$

Definition 6.4.1. The Khatri-Rao product of order $g(\leq n)$ under the decomposition (6.4.1) is defined as

$$N^{(3)} = \bigcup_{(h_1, h_2, \ldots, h_n) \in T} \left(\bigotimes_{i=1}^{n} N_{ih_i} \right)$$

where as in Definition 6.3.1, T denotes a subset of the possible combinations $(h_1, h_2, \ldots, h_n)$ forming an orthogonal array of strength g.

Clearly, $N^{(3)}$ reduces to the ordinary Kronecker product if $g = n$. The following result may be proved along the line of Theorem 6.3.1.

Theorem 6.4.1. If for each i, $h_i (0 \le h_i \le u_i - 1, 1 \le i \le n)$, the design represented by N_{ih_i} be equireplicate with replication number $u_i^{-1} r_i$, then the design $N^{(3)}$ obtained using the method of Khatri-Rao product of order g has OFS with $E_p^x \ge \max_{1 \le i \le n} \{x_i E_{ip}\}$, $0 \le p \le \infty$, for every $x \in \Omega_g^*$. Furthermore, when F^x represents the ith main effect $E_p^x = E_{ip}$.

The observations made in the remark following Theorem 6.3.1 hold in the present set-up as well. Under the conditions of Theorem 6.4.1, the design represented by $N^{(3)}$ has replication number $s \Pi u_i^{-1} r_i$ and block size Πk_i; s being the cardinality of T. Thus as compared to the ordinary Kronecker product, the Khatri-Rao product can achieve a reduction only in the number of replications but not in the block size.

The matrices N_{ih_i} can be obtained easily when each D_i is $(u_i^{-1} r_i)$-resolvable. Then the blocks of D_i may be written as columns in the array z_i as follows:

$$z_i = (z_{i0}^*, z_{i1}^*, \ldots, z_{i,u_i-1}^*)$$

where each symbol is repeated $u_i^{-1} r_i$ times in $z_{ih_i}^*$ $(0 \le h_i \le u_i - 1)$. If N_{ih_i} denotes the incidence matrix of the design with blocks given by the columns of $z_{ih_i}^*$ then clearly the conditions of Theorem 6.4.1 are satisfied.

Example 6.4.1. To construct a $4 \times 6 \times 9$ factorial design, let D_1, D_2, D_3 be varietal designs such that the corresponding arrays z_i are,

$$z_1 = \begin{matrix} 0\,2\,0\,1 \\ 1\,3\,2\,3 \end{matrix}, \quad z_2 = \begin{matrix} 0\,1\,2\,0\,1\,2 \\ 4\,5\,3\,5\,3\,4 \end{matrix}, \quad z_3 = \begin{matrix} 0\,1\,2\,0\,3\,6 \\ 3\,4\,5\,1\,4\,7 \\ 6\,7\,8\,2\,5\,8 \end{matrix}$$

Then $r_1 = r_2 = r_3 = 2$, $k_1 = k_2 = 2$, $k_3 = 3$, with each D_i being 1-resolvable. Hence if z_j be partitioned as $z_j = (z_{j0}^*, z_{j1}^*)$, where $z_{ih_i}^*$ is $k_i \times (b_i/2)$, then the incidence matrices N_{ih_i} of the varietal designs $z_{ih_i}^*$ satisfy the conditions of Theorem 6.4.1 with $u_1 = u_2 = u_3 = 2$. As in Example 6.3.1, the $4 \times 6 \times 9$ factorial design with incidence matrix $N^{(3)}$ formed according to Definition 6.4.1 taking $T = \{(0,0,0),(0,1,1),(1,0,1),(1,1,0)\}$ will have OFS, and Theorem 6.4.1 holds with g = 2. Using (6.3.6), which holds also with the Khatri-Rao product, the A-efficiencies of F_1, F_2, F_3, F_1F_2, F_1F_3, F_2F_3 are 0.6, 0.4286, 0.6667, 0.8347, 0.9,

0.8706 respectively. The design involves blocks of size 12 and 4 replications and is connected.

The methods discussed so far in this chapter may be modified/extended in several directions. Mukerjee (1981a) considered the cyclic product of varietal designs which can control only the main effect efficiencies in the resulting factorial design. These results have an analogy with those in Mukerjee (1982) who considered methods of construction retaining full information on the main effects. Gupta (1983b) considered a combination of the componentwise Kronecker product and the Khatri-Rao product and also a combination of the componentwise Kronecker product and the cyclic product for the construction of factorial designs with OFS, controlling the main effect efficiencies. Mukerjee (1986) indicated the use of Kronecker or Kronecker-type products for generating more complex factorial designs from simpler factorials. For further details regarding these developments, we refer to the original papers.

6.5. Non-equireplicate designs

So far, we have considered only the construction of equireplicate factorial designs. However, there are practical situations where the treatment combinations in a factorial experiment may not be equally expensive. In such situations, the use of equireplicate designs may be either too expensive or wasteful. The use of the Kronecker product for the construction of non-equireplicate factorials will be considered in this section.

As before, for $1 \le i \le n$, let D_i be a varietal block design in m_i treatments and N_i be the incidence matrix of D_i. The designs D_i, $1 \le i \le n$, are no longer assumed to be equireplicate or proper. For $1 \le i \le n$, let b_i be the number of blocks, r_{ij} $(0 \le j \le m_i - 1)$ be the replication numbers and $k_{ij} (0 \le j \le b_i - 1)$ be the block sizes in D_i. Let $\underline{r}_i = (r_{i0}, r_{i1}, \ldots, r_{im_i - 1})'$, $\underline{k}_i = (k_{i0}, k_{i1}, \ldots, k_{ib_i - 1})'$, $R_i = Diag(r_{i0}, r_{i1}, \ldots, r_{im_i - 1})$, $K_i = Diag(k_{i0}, k_{i1}, \ldots, k_{ib_i - 1})$, $1 \le i \le n$. The C-matrix of D_i is then given by, say,

$$C_i = R_i - N_i K_i^{-1} N_i' . \tag{6.5.1}$$

The efficiency with which a treatment contrast with the $m_i \times 1$ co-efficient vector ξ_i is estimated in D_i is defined as

$$E_i(\underline{\xi}_i) = \underline{\xi}_i' R_i^{-1} \underline{\xi}_i / \underline{\xi}_i' C_i^{-} \underline{\xi}_i \text{ if the contrast is estimable in } D_i, \qquad = 0 \quad \text{otherwise}, \tag{6.5.2}$$

where C_i^- is any g-inverse of C_i. Clearly (6.5.2) is based on a comparison of D_i with the corresponding (unblocked) completely randomized design. Note that (6.5.2) remains the same for every choice of the g-inverse C_i^-.

Let D be the $m_1 \times m_2 \times \cdots \times m_n$ Kronecker factorial design with incidence matrix

$$N = \bigotimes_{i=1}^{n} N_i . \tag{6.5.3}$$

The replication numbers and block sizes in D are elements of the vectors $\underline{r} = \underline{r}_1 \otimes \cdots \otimes \underline{r}_n$ and $\underline{k} = \underline{k}_1 \otimes \cdots \otimes \underline{k}_n$ respectively. Note that neither the replication numbers nor the block sizes in D are necessarily equal. Let R and K be diagonal matrices with diagonal elements given by the elements of $\underline{r}$ and $\underline{k}$ respectively. Then by (6.5.1), (6.5.3), the C-matrix of D equals

$$C = R - NK^{-1}N' = \bigotimes_{i=1}^{n} R_i - \bigotimes_{i=1}^{n} (R_i - C_i). \tag{6.5.4}$$

Analogously to (6.5.2), the efficiency with which a treatment contrast $\underline{\xi}'\underline{\tau}$ is estimated in D is defined as

$$E(\underline{\xi}) = \underline{\xi}' R^{-1} \underline{\xi} / \underline{\xi}' C^{-} \underline{\xi} \quad \text{if the contrast is estimable in } D, \qquad = 0 \quad \text{otherwise}, \tag{6.5.5}$$

C^- being any g-inverse of C. It may be seen that in D all main effect contrasts are estimable if and only if each of $D_1, \ldots, D_n$ is connected. Hence it is hereafter assumed that each of $D_1, \ldots, D_n$ is connected. This implies the connectedness of D and the existence of all the matrix inverses used in this section.

The following lemmas will be helpful. The proof of the first lemma is straightforward while that of the second lemma follows essentially along the line of Rao (1973a, pp. 70). As usual, the column space of a matrix A will be denoted by $\mu(A)$.

Lemma 6.5.1. For $1 \leq i \leq n$, let A_i, B_i be n.n.d. matrices such that $A_i - B_i$ is n.n.d. Then

$$\bigotimes_{i=1}^{n} A_i - \bigotimes_{i=1}^{n} B_i$$

is n.n.d.

Lemma 6.5.2. Let A, B be n.n.d. matrices such that $A - B$ is n.n.d. Then $\mu(B) \subseteq \mu(A)$ and for every vector $\underline{\xi} \in \mu(B)$,

$$\underline{\xi}' B^- \underline{\xi} \geq \underline{\xi}' A^- \underline{\xi},$$

where A^-, B^- are any g-inverses of A, B respectively.

For $1 \leq i \leq n$, let $\underline{\xi}_i$ be any $m_i \times 1$ non-null vector satisfying $\underline{\xi}_i' \underline{1}_i = 0$. For any $x = (x_1, \ldots, x_n) \in \Omega$, define the vector

$$\underline{\xi}^x = \bigotimes_{i=1}^{n} \underline{\xi}_i^{x_i}, \tag{6.5.6}$$

where $\underline{\xi}_i^{x_i} = \underline{\xi}_i$ if $x_i = 1$, $= \underline{1}_i$ if $x_i = 0$. Then $\underline{\xi}^{x\prime} \underline{\tau}$ represents a typical contrast belonging to the interaction F^x. We are now in a position to present the counterpart of Theorem 6.2.1 in a non-equireplicate setting.

Theorem 6.5.1. For each $x \in T$, $E(\underline{\xi}^x) \geq \max_{i:x_i=1} E_i(\underline{\xi}_i)$.

Proof. For $1 \leq i \leq n$, let

$$\left.\begin{array}{l} L_i = \bigotimes_{j=1}^{n} L_{ij}\ ,\ W_i = \bigotimes_{j=1}^{n} W_{ij}, \\ \textit{where} \quad L_{ij} = W_{ij} = R_j \text{ if } j \neq i;\ L_{ii} = C_i,\ W_{ii} = R_i - C_i. \end{array}\right\} \tag{6.5.7}$$

Note that for each i,

$$L_i = \bigotimes_{j=1}^{n} R_j - W_i,$$

so that by (6.5.4),

$$C - L_i = W_i - \bigotimes_{j=1}^{n}(R_j - C_j), \tag{6.5.8}$$

which is non-negative definite by (6.5.7) and Lemma 6.5.1.

Consider now any $x = (x_1, \ldots, x_n) \in \Omega$. Without loss of generality, let $x_1 = 1$. Then by (6.5.6), (6.5.7), $\underline{\xi}^x \in \mu(L_1)$. Since by (6.5.8), $C - L_1$ is n.n.d., it follows from Lemma 6.5.2 that

$$\begin{aligned}\underline{\xi}^{x'} C^- \underline{\xi}^x \leq \underline{\xi}^{x'} L_1^- \underline{\xi}^x &= \underline{\xi}^{x'}(C_1^- \otimes R_2^{-1} \otimes \cdots \otimes R_n^{-1})\underline{\xi}^x \\ &= (\underline{\xi}_1' C_1^- \underline{\xi}_1) \times \{\prod_{i=2}^{n}(\underline{\xi}_i^{x_i})' R_i^{-1} \underline{\xi}_i^{x_i}\},\end{aligned} \tag{6.5.9}$$

using (6.5.6), (6.5.7). But by (6.5.6), noting that $x_1 = 1$,

$$\underline{\xi}^{x'} R^{-1} \underline{\xi}^x = (\underline{\xi}_1' R_1^{-1} \underline{\xi}_1) \times \{\prod_{i=2}^{n}(\underline{\xi}_i^{x_i})' R_i^{-1} \underline{\xi}_i^{x_i}\}.$$

Hence by (6.5.2), (6.5.5), (6.5.9), $E(\underline{\xi}^x) \geq E_1(\underline{\xi}_1)$. Similarly, $E(\underline{\xi}^x) \geq E_i(\underline{\xi}_i)$ for every i such that $x_i = 1$. This completes the proof.

As may be seen through examples, very often one gets the satisfying observation that the actual value of $E(\underline{\xi}^x)$ is much higher than the lower bound given in the theorem. Unlike in the equireplicate case, the lower bound may be strict even for the main effects.

Non-equireplicate Kronecker factorial designs may not have OFS in the sense that, as examples reveal, they may not satisfy the necessary and sufficient condition for OFS given by Theorem 4.2.1. Still then, a method for computing a g-inverse of C, which does not require the inversion of large matrices, is available. For $1 \leq i \leq n$, let Δ_i be an $(m_i - 1) \times m_i$ matrix such that $(\rho_i \underline{r}_i, \Delta_i')'$ is orthogonal where $\rho_i = (\underline{r}_i' \underline{r}_i)^{-1/2}$. For $x = (x_1, \ldots, x_n) \in \Omega$, let

$$\Delta^x = \bigotimes_{i=1}^{n} \Delta_i^{x_i},$$

where $\Delta_i^{x_i} = \Delta_i$ if $x_i = 1$, $= \underline{1}_i'$ if $x_i = 0$. Let $\Delta = (\cdots, \Delta^{x'}, \cdots)'_{x \in \Omega}$, e.g., for $n = 2$, $\Delta = (\Delta^{01'}, \Delta^{10'}, \Delta^{11'})'$. Then from (6.5.4) observing that $\Delta^x C \Delta^{y'} = 0$ for every $x, y \in \Omega$, $x \neq y$, it follows that a g-inverse of C is given by

$$C^{-} = \sum_{x \in \Omega} \Delta^{x\prime}(\Delta^{x} C \Delta^{x\prime})^{-1}\Delta^{x}.$$

For each $x \in \Omega$, $\Delta^{x} C \Delta^{x\prime}$ is a square matrix of order $\alpha(x)$. Hence the evaluation of C^{-} requires inversion of matrices of order $\alpha(x)$, $x \in \Omega$.

Before concluding this section, the following open problems in the context of non-equireplicate factorials may be mentioned: (a) role of partial OFS (cf. Chauhan and Dean (1986)), (b) construction in generalized cyclic designs and (c) use of Kronecker-type products for construction in designs of smaller size.

6.6. Designs for multiway heterogeneity elimination

It is possible to extend many of the results in this chapter to a set-up for multiway heterogeneity elimination. For $1 \leq i \leq n$, let D_i be a varietal design for t-way heterogeneity elimination involving m_i treatments, ν_i observations and having a design matrix

$$X_i = [X_{i0}, X_{i1}, \ldots, X_{it}],$$

where X_{i0} is $\nu_i \times m_i$ and X_{ij} is of order $v_i \times b_{ij}$ $(1 \leq j \leq t)$, b_{ij} being the number of classes according to the jth way of heterogeneity elimination. The m_i columns of X_{i0} correspond to the effects of the m_i treatments in D_i while for $1 \leq j \leq t$, the b_{ij} columns of X_{ij} correspond to the effects of the b_{ij} classes according to the jth way of heterogeneity elimination.

The Kronecker product of $D_1, \ldots, D_n$ is an $m_1 \times \cdots \times m_n$ factorial design D, for t-way heterogeneity elimination, involving $\Pi\nu_i$ observations and a design matrix

$$X = [\bigotimes_{i=1}^{n} X_{i0}, \bigotimes_{i=1}^{n} X_{i1}, \ldots, \bigotimes_{i=1}^{n} X_{it}],$$

where the columns of $\bigotimes_{i=1}^{n} X_{i0}$ correspond to the effects of the Πm_i treatment combinations and for $1 \leq j \leq t$, the columns of $\bigotimes_{i=1}^{n} X_{ij}$ correspond to the classes according to the jth way of heterogeneity elimination. Physically, this means that if for $1 \le i \le n$, the treatment j_i occurs in the $(l_{i1}, \ldots, l_{it})th$ "cell" of D_i, then the treatment combination $(j_1, \ldots, j_n)$ occurs in the $((l_{11}, \ldots, l_{n1}), (l_{12}, \ldots, l_{n2}), \ldots, (l_{1t}, \ldots, l_{nt}))$ th "cell" of D.

Example 6.6.1. Let $n = 2$, $t = 2$, $m_1 = 3$, $m_2 = 4$, and D_1, D_2 be row-column designs such that

$$D_1: \begin{array}{ccc} 0 & 1 & 2 \\ 1 & 2 & 0 \end{array} \qquad D_2: \begin{array}{cccc} 0 & 3 & 1 & - \\ 2 & 1 & - & 0 \\ 3 & - & 2 & 1 \\ - & 2 & 0 & 3 \end{array}$$

Then their Kronecker product D is a 3×4 factorial design laid out in 8 rows and 12 columns as shown below:

$$D: \begin{array}{cccccccccccc} 00 & 03 & 01 & - & 10 & 13 & 11 & - & 20 & 23 & 21 & - \\ 02 & 01 & - & 00 & 12 & 11 & - & 10 & 22 & 21 & - & 20 \\ 03 & - & 02 & 01 & 13 & - & 12 & 11 & 23 & - & 22 & 21 \\ - & 02 & 00 & 03 & - & 12 & 10 & 13 & - & 22 & 20 & 23 \\ 10 & 13 & 11 & - & 20 & 23 & 21 & - & 00 & 03 & 01 & - \\ 12 & 11 & - & 10 & 22 & 21 & - & 20 & 02 & 01 & - & 00 \\ 13 & - & 12 & 11 & 23 & - & 22 & 21 & 03 & - & 02 & 01 \\ - & 12 & 10 & 13 & - & 22 & 20 & 23 & - & 02 & 00 & 03 \end{array}$$

It may be noted that the rows and/or columns may be incomplete. Moreover, as in this example, some cells may be left empty.

Just as in the case of block designs, interaction efficiencies in D can be controlled by suitably choosing $D_1, \ldots, D_n$. Thus if $D_1, \ldots, D_n$ be equireplicate then Theorem 6.2.1 holds in a setting for multiway elimination of heterogeneity. On the other hand if $D_1, \ldots, D_n$ are not necessarily equireplicate then Theorem 6.5.1 remains valid under such a setting. However, unlike in the case of block designs, the proofs require the use of projection operators. The relevant details are available in Mukerjee and Sen (1987).

As discussed in Section 6.2, a practical difficulty with the ordinary Kronecker product is that especially with a large number of factors, the Kronecker design may become too large. In the equireplicate case, a variant of the componentwise Kronecker product can be employed to overcome this difficulty and an analogue of

Theorem 6.3.1 can be proved in a set-up for multiway heterogeneity elimination. For the details, we again refer to Mukerjee and Sen (1987).

CHAPTER 7
MORE ON SINGLE REPLICATE FACTORIAL DESIGNS

7.1. Introduction

Ever since Fisher (1935) popularized factorial designs, the construction of such designs in a single replicate has received a lot of attention. In fact, most of the classical results on factorial designs are centered around single replicate designs. In many practical situations, especially when the number of factors is large, it may be too expensive to have more than a single replicate. Even in situations where the available resources allow the use of several replicates, it may sometimes be convenient to start from single replicate designs and then to combine them to construct resolvable multireplicate designs whose properties, in terms of interaction efficiencies, can be analyzed from those of the constituent single replicate designs.

Single replicate factorial experiments in generalized cyclic designs were considered in Chapter 5. In this chapter, we propose to present some further details following Voss (1986) and Voss and Dean (1987) who explored the interrelationship among the various construction procedures and attempted to integrate them.

7.2. General classical designs

For the case of symmetric, prime powered factorial block designs, Bose (1947) developed a general theory incorporating the previous results by Bose and Kishen (1940). Generalizations of this classical method to asymmetric factorial experiments are also available in the literature. Factorial experiments, where each factor occurs at one of two prime numbers of levels were considered by White and Hultquist (1965). The results of White and Hultquist were generalized by Raktoe (1969, 1970) to factorial experiments for which each factor occurs at one of several prime-powered numbers of levels. Further results were reported by Worthley and Banerjee (1974) and Sihota and Banerjee (1981). All but the latest of these were reviewed and discussed by Raktoe, Rayner and Chalton (1978). Each of the above generalizations of the classical method, with the exception of a generalization of Raktoe (1970), is equivalent to applying the classical method of Bose (1947) separately to each symmetric subexperiment. Voss (1986) investigated the relationship among the various methods and showed that, barring that due to Raktoe (1970) for experiments involving factors with numbers of levels being different powers of the same prime, each produce classes of designs which form subsets of a more general class namely the general classical class of designs.

Consider an $m_1^{n(1)} \times m_2^{n(2)} \times \cdots \times m_g^{n(g)}$ factorial block design involving $n = \sum_{j=1}^{g} n(j)$ factors with factor F_{jh} occurring at m_j levels, $1 \le h \le n(j)$, $1 \le j \le g$. Denote a treatment combination by $i = (i_1, i_2, \ldots, i_g)$, where $i_j = (i_{j1}, i_{j2}, \ldots, i_{jn(j)})$ denotes the subtreatment combination for the jth symmetric subexperiment involving the $n(j)$ factors at m_j levels. The set of treatment combinations forms a group G with the group operation addition componentwise modulo m_j defined as follows:

$$(i_{11}, \ldots, i_{gn(g)}) + (i'_{11}, \ldots, i'_{gn(g)}) = (i^*_{11}, \ldots, i^*_{gn(g)}), \tag{7.2.1}$$

where $i^*_{jh} = i_{jh} + i'_{jh} \; (mod \; m_j)$, $1 \le h \le n(j)$, $1 \le j \le g$.

Remark. It is possible to represent G as the direct sum $\bigoplus_{j=1}^{g} \bigoplus_{h=1}^{n(j)} Z(m_j)$, where $Z(m_j)$ is the ring of integers modulo m_j. If m_j is a prime-power, then the ring $Z(m_j)$ may be replaced by the Galois field $GF(m_j)$ and the group operation of G is addition componentwise $GF(m_j)$.

For $1 \le j \le g$, let $G_j = \{i_j\colon\ i_j = (i_{j1}, \ldots, i_{jn(j)}),\ 0 \le i_{jh} \le m_j - 1\ (1 \le h \le n(j))\}$. Then G_j forms a group under componentwise addition modulo m_j.

Definition 7.2.1 (Voss, 1986). A general classical design is a factorial block design such that for fixed subgroups $A_j \subset G_j$ $(1 \le j \le g)$, the principal block (that is, the block containing the treatment combination (0, 0, ..., 0)) is the subgroup

$$\begin{aligned} B_0 = \{ i\colon\ & i = (i_1, \ldots, i_g) \in G,\ i_j = (i_{j1}, \ldots, i_{jn(j)}) \in G_j \\ & \text{and } \sum_{h=1}^{n(j)} a_{jh} i_{jh} = 0 \ (mod\ m_j) \\ & \text{for } every\ a_j = (a_{j1}, \ldots, a_{jn(j)}) \in A_j,\ 1 \le j \le g \}, \end{aligned}$$

and the other blocks are obtained as cosets of B_0 under componentwise addition modulo m_j. For m_j a prime power, $GF(m_j)$ may be used instead of $Z(m_j)$.

The general classical method, as described above, is equivalent to applying the classical method of Bose (1947) separately within each symmetric $m_j^{n(j)}$ experiment, using either $Z(m_j)$ or $GF(m_j)$ for each.

Example 7.2.1. Consider a $3^2 \times 2$ experiment with factors F_{11} and F_{12} each at three levels and factor F_{21} at two levels. A general classical design is determined by the subgroups $A_1 = \{(0,0),(1,1),(2,2)\}$ and $A_2 = \{(0),(1)\}$. Since (1,1) generates A_1 and (1) generates A_2, by Definition 7.2.1,

$$B_0 = \{i: \; i = (i_{11},i_{12},i_{21}), \; i_{11} + i_{12} = 0 \; (mod\ 3), \quad i_{21} = 0 \; (mod\ 2)\}$$
$$= \{(0,0,0),(1,2,0),(2,1,0)\}.$$

The other blocks, which are cosets of B_0 in G, are given by $\{(0,0,1),(1,2,1),(2,1,1)\}$, $\{(0,1,0),(1,0,0),(2,2,0)\}$, $\{(0,1,1),(1,0,1),(2,2,1)\}$, $\{(0,2,0),(1,1,0),(2,0,0)\}$ and $\{(0,2,1),(1,1,1),(2,0,1)\}$. It can be seen (cf. Theorem 7.2.2 below) that in this design 1 degree of freedom (d.f.) belonging to main effect F_{21} and 2 d.f.'s belonging to each of interactions $F_{11}F_{12}$ and $F_{11}F_{12}F_{21}$ are not estimable.

As Voss (1986) had shown, the class of general classical designs includes the classes of designs obtained by the methods of White and Hultquist (1965), Raktoe (1969), Worthley and Banerjee (1974), Sihota and Banerjee (1981) and the method of Raktoe (1970) restricted to experiments where the numbers of levels of different factors are either equal or powers of different primes. We denote an algebra of order s by $A(s)$ and an element of $A(s)$ by $a(s)$.

Assuming the above methods apply in an $m_1^{n(1)} \times m_2^{n(2)} \times \cdots \times m_g^{n(g)}$ experiment, the principal block is of the form

$$B_0^* = \{i: \; \sum_{j=1}^{g} \sum_{h=1}^{n(j)} a_{jhu}(m_j) i_{jh}(m_j) = 0(m), \; 1 \le u \le d\}, \tag{7.2.2}$$

for some positive integer d, where $m = \prod_{j=1}^{g} m_j$. For addition of elements from different algebras to be well-defined in (7.2.2), each of the above authors defines mappings $\sigma_j\colon A(m_j) \to A(m)$ such that the subalgebras $A(m_j)$, $1 \le j \le g$, are either embedded in a superalgebra $A(m)$ or elements of the same subalgebra are required to be combined within the subalgebra, then mapped into the superalgebra;

then operations are defined over the superalgebra. The mappings σ_j are so defined that each element $a(m)$ has a unique decomposition $a(m) = \sum_{j=1}^{g} \sigma_j(a_j(m_j))$. Then by (7.2.2)

$$\begin{aligned} B_0^* &= \{i: \sum_{j=1}^{g} \sigma_j(\sum_{h=1}^{n(j)} a_{jhu}(m_j) i_{jh}(m_j)) = 0(m), \ 1 \le u \le d\} \\ &= \{i: \sum_{j=1}^{g} \sigma_j(\sum_{h=1}^{n(j)} a_{jhu}(m_j) i_{jh}(m_j)) = \sum_{j=1}^{g} \sigma_j(0(m_j)), \ 1 \le u \le d\} \\ &= \{i: \sum_{h=1}^{n(j)} a_{jhu}(m_j) i_{jh}(m_j) = 0(m_j), \ 1 \le j \le g, \ 1 \le u \le d\}. \end{aligned} \tag{7.2.3}$$

From Definition 7.2.1 and (7.2.3), it follows that, when applicable, the methods due to the above authors generate classes of designs which form subclasses of the class of general classical designs.

As Voss (1986) remarks, the class of general classical designs does not contain the class of designs generated by the method of Raktoe (1970) for experiments where the numbers of levels of different factors are powers of the same prime since the elements of $GF(p^n)$ do not yield the necessary decomposition into a sum of elements of subfields.

In the rest of this section, we consider further properties of general classical designs. To that effect, with reference to Definition 7.2.1, for $1 \le j \le g$, let N_j be the incidence matrix of a single replicate $m_j^{n(j)}$ symmetric factorial design with the principal block given by

$$B_{0j} = \{i_j: \ i_j = (i_{j1}, \ldots, i_{jn(j)}) \in G_j \ \text{and} \ \sum_{h=1}^{n(j)} a_{jh} i_{jh} = 0 \ (mod \ m_j),$$
$$\text{for every} \ \ a_j = (a_{j1}, \ldots, a_{jn(j)}) \in A_j\},$$

and the other blocks obtained as cosets of B_{0j} in G_j under componentwise addition modulo m_j (as before, for m_j a prime power, $GF(m_j)$ may be used instead of $Z(m_j)$). Also, let N be the incidence matrix of an $m_1^{n(1)} \times \cdots \times m_g^{n(g)}$ general classical design as given by Definition 7.2.1. Then, as shown by Voss (1986),

$$N = \bigotimes_{j=1}^{g} N_j. \tag{7.2.4}$$

If m_j is a prime power and field arithmetic is used to construct N_j, then following

Bose (1947) it may be seen that $N_j N_j'$ has structure K. Otherwise, N_j is a GC design and, as in Chapter 5, $N_j N_j'$ has structure K. Hence by (7.2.4),

$$NN' = \bigotimes_{j=1}^{g} N_j N_j'$$

has structure K and, applying Theorem 4.3.2, the following result is evident.

<u>Theorem 7.2.1 (Voss, 1986).</u> The class of general classical designs is regular and has OFS.

The next result deals with the efficiency of a general classical design N for the estimation of treatment contrasts. It is easy to see that in a single replicate design the best linear unbiased estimator (BLUE) of an estimable treatment contrast is given by the corresponding observational contrast and, as such, all estimable treatment contrasts are estimated with efficiency 1, where, as usual, efficiency is measured relative to the corresponding complete block design. In other words, the efficiency of a single replicate design with respect to a treatment contrast equals 1 or 0, according as the contrast is estimable or not.

Now for $1 \le j \le g$, let $\underset{\sim}{\xi}_j$ be an $m_j^{n(j)} \times 1$ non-null vector such that $\underset{\sim}{\xi}_j' \underset{\sim}{\epsilon}_j = 0$, where the $m_j^{n(j)} \times 1$ vector $\underset{\sim}{\epsilon}_j$ has all elements unity. For $x = (x_1, \ldots, x_g) \neq 0$, let

$$\underset{\sim}{\xi}^x = \bigotimes_{j=1}^{g} \underset{\sim}{\xi}_j^{x_j}, \tag{7.2.5}$$

where $\underset{\sim}{\xi}_j^{x_j} = \underset{\sim}{\xi}_j$ if $x_j = 1$, $= \underset{\sim}{\epsilon}_j$ if $x_j = 0$. Denoting the vector of treatment effects in a general classical design N by $\underset{\sim}{\tau}$, it follows that $\underset{\sim}{\xi}^{x\prime} \underset{\sim}{\tau}$ represents a treatment contrast in N. Let $E(\underset{\sim}{\xi}^x)$ be the efficiency of N with respect to $\underset{\sim}{\xi}^{x\prime} \underset{\sim}{\tau}$. Similarly, for $1 \le j \le g$, let $E_j(\underset{\sim}{\xi}_j)$ be the efficiency of N_j with respect to a treatment contrast with co-efficient vector $\underset{\sim}{\xi}_j$. Then one obtains the following result, which was stated by Voss (1986) in a slightly different way:

<u>Theorem 7.2.2.</u> $E(\underset{\sim}{\xi}^x) = \max_{j:x_j=1} E_j(\underset{\sim}{\xi}_j) = 1 - \prod_{j:x_j=1} \{1 - E_j(\underset{\sim}{\xi}_j)\}$.

Proof. It is easy to see (cf. Dean (1978)) that the contrast $\underline{\xi}^{x\prime}\underline{\tau}$ is estimable in the single replicate design N if and only if $N'\underline{\xi}^{x} = \underline{0}$. Similarly, for each j, the treatment contrast with co-efficient vector $\underline{\xi}_j$ is estimable in the design N_j if and only if $N_j'\underline{\xi}_j = \underline{0}$. Since by (7.2.4), (7.2.5)

$$N'\underline{\xi}^{x} = \bigotimes_{j=1}^{g}(N_j'\underline{\xi}_j^{x_j}),$$

it follows that $\underline{\xi}^{x\prime}\underline{\tau}$ is estimable in N if and only if for at least one j, with $x_j = 1$, the treatment contrast with co-efficient vector $\underline{\xi}_j$ is estimable in N_j. Hence recalling that in a single replicate design the efficiency with respect to a treatment contrast equals 1 or 0 according as the contrast is estimable or not, the result follows immediately.

Recall from Chapter 5 that a single replicate generalized cyclic design is a design whose principal block forms a subgroup of G, the group of all treatment combinations, the other blocks being the cosets of the principal block. Hence from Definition 7.2.1, the following result is evident.

Theorem 7.2.3 (Voss and Dean, 1987). The class of general classical designs using the rings $Z(m_j)$, $1 \leq j \leq g$, forms a subclass of the generalized cyclic designs.

Further results on the relationship between the classes of general classical and generalized cyclic designs will be discussed in the next section.

7.3. Bilinear classical designs

Patterson (1965, 1976) presented an algorithm, called DSIGN, for the computer generation of factorial designs. The algorithm is available for any of the simple block structures defined by Nelder (1965). It generates a large class of designs including the class of single replicate generalized cyclic designs (see Patterson (1976), Bailey (1977)). Bailey, Gilchrist and Patterson (1977) and Bailey (1977) described a method of identifying confounding patterns in a large subclass of designs generated by DSIGN. Patterson and Bailey (1978) explained through examples the use of the algorithm.

Further related results were reported by El Mossadeq, Kobilinsky and Collombier (1985). An alternative but related method was given by Collings (1984). A concise, informative account of the DSIGN algorithm, in the context of single replicate block designs, is available in Street and Street (1987).

As indicated by Voss and Dean (1987), the annihilator method of Bailey (1977) is the most general of the DSIGN related construction procedures mentioned above. When restricted to single replicate block designs, the resulting class of designs will be referred to as bilinear classical designs. As in Section 7.2, consider an $m_1^{n(1)} \times m_2^{n(2)} \times \cdots \times m_g^{n(g)}$ factorial experiment and let G be the group of the treatment combinations with group operation defined by (7.2.1). For each pair of elements i, $a \in G$, define the integer [a,i] as

$$[a,i] = \sum_{j=1}^{g} \sum_{h=1}^{n(j)} a_{jh} i_{jh} \gamma m_j^{-1} \quad (mod\ \gamma), \tag{7.3.1}$$

where γ is the lowest common multiple of $m_1, m_2, \ldots, m_g$.

Definition 7.3.1 (Bailey, 1977). A bilinear classical design is a design such that for a fixed subgroup A of G, the principal block is the subgroup $B_0 = \{i: i \in G, [a,i] = 0$ for all $a \in A\}$, and the other blocks are cosets of B_0.

Following Bailey, Gilchrist and Patterson (1977), each $a = (a_1, a_2, \ldots, a_g)$, where $a_j = (a_{j1}, a_{j2}, \ldots, a_{jn(j)})$ represents one d.f. belonging to the interaction among those factors for which $a_{jh} \neq 0$. The set A represents the subgroup of d.f.'s which are confounded with blocks.

Example 7.3.1. Consider a $2^2 \times 3^2$ bilinear classical design generated by the subgroup $A = \{(0,0,0,0),(1,1,1,2),(0,0,2,1),(1,1,0,0),(0,0,1,2),(1,1,2,1)\}$ of G. Here $\gamma = 6$ and the principal block, which consists of the treatment combinations satisfying $3(i_{11}+i_{12}) + 2(i_{21}+2i_{22}) = 0 \pmod 6$, is given by the subgroup $B_0 = \{(0,0,0,0),(0,0,1,1),(0,0,2,2),(1,1,0,0),(1,1,1,1),(1,1,2,2)\}$. The other blocks are the cosets of B_0 in G. Thus one obtains a $2^2 \times 3^2$ single replicate factorial design in 6 blocks each of size 6. It can be seen that in this design 1 d.f. belonging to interaction $F_{11}F_{12}$ and 2 d.f.'s belonging to each of interactions $F_{21}F_{22}$ and

$F_{11}F_{12}F_{21}F_{22}$ are not estimable.

Theorem 7.3.1. (Voss and Dean, 1987). The classes of single replicate generalized cyclic designs and bilinear classical designs are identical.

Proof. By Definition 7.3.1, every bilinear classical design is also a generalized cyclic design. Conversely, for a generalized cyclic design with principal block B_0, let $B_0^0 = \{a\colon\ a \in G,\ [a,i] = 0 \text{ for all } i \in B_0\}$ be the annihilator of B_0. Since B_0 is a subgroup of G, the annihilator of B_0^0, namely $(B_0^0)^0 = \{i\colon\ i \in G,\ [a,i] = 0$ for all $a \in B_0^0\}$, is B_0 itself. Hence $A = B_0^0$ generates the bilinear classical design with principal block B_0.

It was seen in Theorem 7.2.3 that the class of general classical designs form a subclass of the generalized cyclic designs. It can be shown using Theorem 7.2.2. that for the general classical method the smallest block size possible in an $m_1^{n(1)} \times m_2^{n(2)} \times \cdots \times m_g^{n(g)}$ experiment keeping all main effect contrasts estimable is $\prod_{j=1}^{g} m_j$. The generalized cyclic design generated by (1,1,...,1) has all main effect contrasts estimable and a block size given by the lowest common multiple of $m_1, m_2, \ldots, m_g$ (see Section 5.4). Hence, as Voss and Dean (1987) point out, if $m_1, m_2, \ldots, m_g$ are not all relatively prime then the general classical designs form a proper subclass of the generalized cyclic designs. However, if $g = 1$ or $m_1, m_2, .., m_g$ are relatively prime, then the two classes of designs are identical. This is shown by the next two theorems.

Theorem 7.3.2 (Voss and Dean, 1987). For an $m_1^{n(1)} \times m_2^{n(2)} \times \cdots \times m_g^{n(g)}$ factorial experiment, the generalized cyclic (bilinear classical) designs form a subclass of the general classical designs if $m_1, m_2, \ldots, m_g$ are relatively prime.

Proof. By Theorem 7.3.1, the principal block of a generalized cyclic design can be expressed as $B_0 = \{i\colon\ i \in G,\ [a,i] = 0 \text{ for all } a \in A\}$ for some subgroup A of G. Here, by (7.3.1), $[a,i] = \gamma \sum_{j=1}^{g} m_j^{-1} s_j$, where $s_j = \sum_{h=1}^{n(j)} a_{jh} i_{jh}$ and $\gamma = \prod_{j=1}^{g} m_j$ as $m_1, m_2, \ldots, m_g$ are relatively prime. Consider $i \in B_0$. Since $[a,i] = 0 \ (mod\ \gamma)$,

$\sum_{j=1}^{g} m_j^{-1} s_j$ is an integer. For $m_1, m_2, \ldots, m_g$ relatively prime, it can then be shown that $s_j = q_j m_j$ for some integer q_j, i.e. $s_j = \sum_{h=1}^{n(j)} a_{jh} i_{jh} = 0 \ (mod\ m_j)$, $1 \leq j \leq g$. Writing $a = (a_1, a_2, \ldots, a_g)$, where $a_j = (a_{j1}, \ldots, a_{jn(j)})$, $1 \leq j \leq g$, let $A_j = \{a_j\colon\ a = (a_1, a_2, \ldots, a_g) \in A\}$. Then clearly,

$$B_0 \subset B_{00} = \{i\colon\ i \in G,\ \sum_{h=1}^{n(j)} a_{jh} i_{jh} = 0 \ (mod\ m_j) \text{ for } all\ \ a_j \in A_j, 1 \leq j \leq g\}.$$

Also if $i \in B_{00}$, then m_j divides s_j, so that $[a, i] = \gamma \sum_{j=1}^{g} m_j^{-1} s_j = 0 \ (mod\ \nu)$, and $i \in B_0$. Therefore, $B_0 = B_{00}$, which is the principal block of a general classical design - the other blocks are cosets.

The following is an immediate consequence of Theorems 7.2.3, 7.3.1 and 7.3.2.

Theorem 7.3.3 (Voss and Dean, 1987). The classes of generalized cyclic, bilinear classical and general classical designs are equivalent using the rings $Z(m_j)$, $1 \leq j \leq g$, when $m_1, m_2, \ldots, m_g$ are relatively prime.

As Voss and Dean (1987) remark, for any m_j which is a prime-power, one could use $GF(m_j)$ in place of $Z(m_j)$ in the construction of both generalized cyclic and general classical designs in which case the resulting classes are still equivalent.

7.4. Concluding remarks.

The results presented in this chapter exhibit that the seemingly different construction procedures for single replicate factorials generate, in most cases, equivalent classes of designs. In certain experimental settings, the use of pseudofactors does not enlarge this common class of designs.

The use of pseudofactors has been considered by several authors - see e.g. Yates (1937), Kempthorne (1952, p. 343), Patterson (1965), Giovagnoli (1977), Collings (1984), and Voss (1988). It was stated by Giovagnoli (1977) that the use of prime-leveled pseudofactors gives the largest class of designs. If prime-leveled pseudofactors are used then by Theorem 7.3.3, the generalized cyclic, bilinear classical and general classical methods generate the same class of designs, namely, the pseudofactor class of designs.

For an $m_1^{n(1)} \times m_2^{n(2)} \times \cdots \times m_g^{n(g)}$ experiment, where none of $m_1, m_2, \ldots, m_g$ has a prime-powered divisor, it was shown by Voss and Dean (1987) that the classes of generalized cyclic, bilinear classical and general classical designs are each d.f. equivalent to the pseudofactor class in the sense of producing designs which confound the same number of d.f.('s) from the same interaction. As demonstrated by Voss and Dean (1987), with reference to an example from Giovagnoli (1977), this d.f. equivalence may not hold if any m_j has a prime-powered divisor. Further related results based on Sylow subgroups, under more general block structures, are available in Bailey (1985b).

CHAPTER 8
FURTHER DEVELOPMENTS

8.1. Introduction

In Chapters 5 and 6, two systematic approaches were described for the construction of factorial designs with OFS controlling the interaction efficiences. It is also possible to start with a $q_1 \times q_2 \times \cdots \times q_n$ factorial design, d_0, and then to construct an $m_1 \times m_2 \times \cdots \times m_n$ factorial design, d, by deletion or merging of treatment labels in d_0. Even though the resulting design d may not have OFS, it is reasonable to anticipate that the properties of d_0 in terms of the estimation of the interaction contrasts will influence those of d. These aspects will be taken up in the next two sections. In Section 8.4, some efficiency and admissibility results, in the context of factorial designs in general, are considered. Finally, in the concluding section we briefly indicate possible applications of the calculus for factorial arrangements to some other areas.

8.2. Deletion designs

Das (1960) gave a method of obtaining balanced designs with OFS for asymmetrical factorial experiments by deleting certain treatment combinations from a suitably constructed symmetric factorial. A similar technique was previously used by Kishen and Srivastava (1959) in constructing designs for asymmetrical factorial experiments by truncating $EG(m,s)$ suitably. Sardana and Das (1965) gave some further designs using this technique.

Recently Voss (1986) considered another type of deletion designs in a single replicate setting. Construct a $q_1 \times q_2 \times \cdots \times q_n$ single replicate block design, d_0, using any procedure available in the literature (e.g., those discussed in Chapter 7). Then $a_i = q_i - m_i$ levels of the ith factor of d_0 are selected and those treatment combinations from d_0 are deleted in which the ith factor occurs at one of these a_i levels. The design, d, so obtained is clearly a single replicate $m_1 \times m_2 \times \cdots \times m_n$ factorial which will be referred to as a deletion design. Voss (1986) considered first-order deletion designs where levels from only one factor of the preliminary design d_0 are deleted. It is assumed, without loss of generality, that levels are deleted from the first factor of design d_0.

Let $\underline{\tau}$ and $\underline{\tau}_0$ denote respectively the column vectors of lexicographically arranged treatment parameters in d and d_0. The ith factor in d and d_0 will be denoted by F_i and F_{i0}, respectively, $1 \le i \le n$. For a treatment contrast $\underline{\xi}'\underline{\tau}$ in d, let $\underline{\xi}_0'\underline{\tau}_0$ be the corresponding contrast in d_0. That is, for a 2×3 first-order deletion design d obtained from a 3×3 factorial design d_0, if $\underline{\xi} = (\xi_{00},\xi_{01},\xi_{02},\xi_{10},\xi_{11},\xi_{12})'$, then $\underline{\xi}_0 = (\xi_{00},\xi_{01},\xi_{02},\xi_{10},\xi_{11},\xi_{12},0,0,0)'$. The following lemma is evident:

Lemma 8.2.1. In a single replicate setting, $\underline{\xi}'\underline{\tau}$ is estimable in d if and only if $\underline{\xi}_0'\underline{\tau}_0$ is estimable in d_0.

Lemma 8.2.2. If $\underline{\xi}'\underline{\tau}$ corresponds to the interaction $F_1F_2^{x_2} \cdots F_n^{x_n}$, then $\underline{\xi}_0'\underline{\tau}_0$ corresponds to the interaction $F_{10}F_{20}^{x_2} \cdots F_{n0}^{x_n}$, $x_i = 0$ *or* 1, $2 \le i \le n$.

Proof. Note that $\underline{\xi} = \underline{\xi}_1 \otimes \underline{\xi}_2^{x_2} \otimes \cdots \otimes \underline{\xi}_n^{x_n}$ where $\underline{\xi}_j^{x_j}$ is an $m_j \times 1$ contrast vector if $x_j = 1$. Otherwise $\underline{\xi}_j^{x_j}$ is proportional to a column vector of ones of appropriate order. Without loss of generality, the corresponding contrast in d_0 can be written as $\underline{\xi}_0 = (\underline{\xi}_1', \underline{0}')' \otimes \underline{\xi}_2^{x_2} \otimes \cdots \otimes \underline{\xi}_n^{x_n}$ which clearly belongs to the interaction $F_{10}F_{20}^{x_2} \cdots F_{n0}^{x_n}$.

Theorem 8.2.1 follows from Lemmas 8.2.1 and 8.2.2.

Theorem 8.2.1. (Voss 1986). If all contrasts belonging to the interaction $F_{10}F_{20}^{x_2} \cdots F_{n0}^{x_n}$ are estimable in d_0 then all contrasts belonging to $F_1F_2^{x_2} \cdots F_n^{x_n}$ are estimable in the first order deletion design d, $x_i = 0$ or 1, $2 \le i \le n$.

Theorem 8.2.2. (Voss, 1986). If all contrasts belonging to $F_{20}^{x_2}F_{30}^{x_3} \cdots F_{n0}^{x_n}$ and $F_{10}F_{20}^{x_2} \cdots F_{n0}^{x_n}$ are estimable in d_0 then all contrasts belonging to $F_2^{x_2}F_3^{x_3} \cdots F_n^{x_n}$ are estimable in d, $x_i = 0$ or 1, $2 \le i \le n$.

Proof. Let $\underline{1}_s$ and $\underline{0}_s$ denote column vectors of $1's$ and $0's$ respectively, both of size s. Then,

$$(\underline{1}'_{m_1}, \underline{0}'_{a_1}) = [m_1/(m_1+a_1)]\underline{1}'_{m_1+a_1} + [(a_1/(m_1+a_1))\underline{1}'_{m_1}, (-m_1/(m_1+a_1))\underline{1}'_{a_1}]$$
$$= \underline{p}'_1 + \underline{p}'_2, \text{ say.}$$

The result follows noting that $\underline{p}_2$ is a contrast vector and using Lemma 8.2.1.

Example 8.2.1 (Voss, 1986). Construct a 3^3 preliminary design d_0 by the classical method of Bose (1947) where principal block is a subgroup (of the group of all treatment combinations) given by {000,012,021,102,111,120,201,210,222} and the other two blocks are cosets thereof (vide Chapter 7). It can be seen that 2 d.f.'s belonging to the interaction $F_{10}F_{20}F_{30}$ are not estimable in d_0. Delete from d_0 those treatments where F_{10} occurs at the highest level to obtain a design d for a 2×3^2 experiment.

Treatment combinations

Block	Kept						Deleted		
1	000	012	021	102	111	120	201	210	222
2	001	010	022	100	112	121	202	211	220
3	002	011	020	101	110	122	200	212	221

Applying Theorems 8.2.1, 8.2.2, it may be seen that in d all contrasts belonging to the main effects and two-factor interactions, except F_2F_3 are estimable.

Voss (1986) also considered the application of the deletion technique to generalized cyclic designs such that under the absence of higher-order interactions, the resulting designs are efficient for the lower-order ones.

8.3. Merging of treatments

The technique of merging of treatments in a given design was indicated by Pearce (1971) and studied by Puri and Nigam (1975, 1976, 1978, 1983). Puri and Nigam (1976, 1978) considered merging of treatments in factorial designs which are balanced with OFS while their other papers were related to varietal designs. More generally, Gupta and Mukerjee (1989) applied the technique to equireplicate factorial designs which have OFS but are not necessarily balanced. The technique is capable of producing non-equireplicate designs where the interaction efficiencies can be controlled by suitably choosing the initial design.

Let d_0 be an initial equireplicate design having common treatment replication number r and v_0 treatments denoted by $0,1,\ldots,v_0-1$, which are classified into v mutually exclusive and exhaustive non-empty subsets. Without loss of generality let these subsets be $\{0,1,\ldots,t_1-1\}$, $\{t_1,t_1+1,\ldots,t_1+t_2-1\}$, ..., $\{t_1+t_2+\cdots+t_{v-1},t_1+t_2+\cdots+t_{v-1}+1,\ldots,t_1+t_2+\cdots+t_v-1\}$, where $\sum_{i=1}^{v} t_i = v_0$ and $t_1, t_2, \ldots, t_v \geq 1$. Now construct a new design d from d_0 by replacing the ith group of treatments by a single treatment i ($1 \leq i \leq v$). Let C and C_0 denote the coefficient matrices of intrablock reduced normal equations for designs d and d_0 respectively, and

$$H = \begin{bmatrix} \underline{1}'_{t_1} & \underline{0}' & \cdots & \underline{0}' \\ \underline{0}' & \underline{1}'_{t_2} & \cdots & \underline{0}' \\ \cdot & \cdot & \cdots & \cdot \\ \cdot & \cdot & \cdots & \cdot \\ \cdot & \cdot & \cdots & \cdot \\ \underline{0}' & \underline{0}' & \cdots & \underline{1}'_{t_v} \end{bmatrix},$$

$$B = HH' = diag\{t_1, \ldots, t_v\}, \ \underline{b} = (t_1, \ldots, t_v)' , \tag{8.3.1}$$

where $\underline{1}_p$ is the vector of 1's of size p and H will be referred to as a collapsing matrix. Then it is easy to see that $C = HC_0H'$ and that the replication vector in d is given by $r\underline{b}$. The design d_0 is assumed to be connected, which also implies the connectedness of d. As before, let $\underline{\tau}$ and $\underline{\tau}_0$ denote the column vectors of treatment parameters in designs d and d_0 respectively. Then analogously to (6.5.5), the efficiency of d with respect to the estimation of a typical treatment contrast $\underline{\xi}'\underline{\tau}$ is defined as,

$$E_d(\underline{\xi}) = r^{-1}\underline{\xi}' B^{-1}\underline{\xi}/\underline{\xi}' C^{-}\underline{\xi}. \tag{8.3.2}$$

Similarly, the efficiency of d_0 with respect to the estimation of a treatment contrast $\underline{a}'\underline{\tau}_0$ is defined as,

$$E_{d_0}(\underline{a}) = r^{-1}\underline{a}'\underline{a}/\underline{a}'C_0^-\underline{a}. \tag{8.3.3}$$

Corresponding to a treatment contrast $\underline{\xi}'\underline{\tau}$ in d, consider a treatment contrast $\underline{\xi}'\underline{B}^{-1}H\underline{\tau}_0$ in d_0.

<u>Theorem 8.3.1 (Gupta and Mukerjee (1989).</u> $E_d(\underline{\xi}) \geq E_{d_0}(H'B^{-1}\underline{\xi})$.

<u>Proof.</u> Since d_0 is connected, we can write $\underline{\xi}'B^{-1}H = \underline{u}'C_0$ for some vector $\underline{u}$. Hence by (8.3.1) $\underline{\xi}' = \underline{u}'C_0H'$. Also the non-negative definite matrix C_0 may be expressed as $C_0 = MM'$ for some matrix M. Hence using $C = HC_0H'$,

$$\begin{aligned}\underline{\xi}'B^{-1}HC_0^-H'B^{-1}\underline{\xi} - \underline{\xi}'C^-\underline{\xi} &= \underline{u}'C_0\underline{u} - \underline{u}'C_0H'(HC_0H')^-HC_0\underline{u} \\ &= \underline{u}'M[I-M'H'(HMM'H')^-HM]M'\underline{u} \geq 0,\end{aligned} \tag{8.3.4}$$

observing the idempotence of the matrix within squared brackets. The result now follows by (8.3.1) - (8.3.4) taking $\underline{a} = H'B^{-1}\underline{\xi}$ in (8.3.3).

The above theorem implies that efficiency factors of treatment contrasts in the derived design can be controlled by choosing an appropriate initial design. Also, if d_0 is not variance-balanced then the inequality given in Theorem 8.3.1 may be strict as can be shown by examples.

Turning to the factorial set-up, let d_0 be an equireplicate design with OFS for an n-factor experiment, the jth factor F_{j0} being at q_j levels, $1 \leq j \leq n$. Let the q_j levels of F_{j0} in d_0 be collapsed into m_j levels $(2 \leq m_j \leq q_j,\ 1 \leq j \leq n)$, and let analogously to H, the associated collapsing matrix be denoted by H_j with $B_j = H_jH_j'$. Then the collapsing matrix for the entire set of treatment combinations may be seen to be $H = \bigotimes_{j=1}^{n} H_j$. Let d be the $m_1 \times m_2 \times \cdots \times m_n$ factorial design obtained from d_0 through such merging of treatments, $\underline{1}_j$ and $\underline{1}_{j0}$ denote column vectors of sizes m_j and q_j respectively and $\underline{\xi}_j$ denote an $m_j \times 1$ contrast vector $(1 \leq j \leq n)$. A typical contrast belonging to the interaction $F_1^{x_1}F_2^{x_2} \cdots F_n^{x_n}$ in d $(x_j = 0$ *or* $1)$ is then given by $\underline{\xi}^{x\prime}\underline{\tau} = (\bigotimes_{j=1}^{n}\xi_j^{x_j})'\underline{\tau}$, where

$x = (x_1, x_2, \ldots, x_n)$, $x_i = 0$ or 1, and $\underline{\xi}_i^{x_i} = \underline{\xi}_i$ if $x_i = 1$, $= \underline{1}_i$ if $x_i = 0$. Then by Theorem 8.3.1,

$$E_d(\underline{\xi}^x) \geq E_{d_0}(H' B^{-1} \underline{\xi}^x) \tag{8.3.5}$$

where $B = \bigotimes_{j=1}^{n} B_j$.

As in Chapter 2, let Ω be the collection of all n-component non-null (0,1)-vectors. For any $x \in \Omega$, let $\Omega^*(x) = \{y = (y_1, y_2, \ldots, y_n)$: $y \in \Omega$, $y_i \geq x_i$, $1 \leq i \leq n\}$. For $1 \leq i \leq n$, let I_i and I_{i0} denote identity matrices of orders m_i and q_i respectively, and as in (2.2.2),

$$M_i^{y_i} = \begin{cases} I_{i0} - q_i^{-1} \underline{1}_{i0} \underline{1}_{i0}' & \text{if } y_i = 1 \\ q_i^{-1} \underline{1}_{i0} \underline{1}_{i0}' & \text{if } y_i = 0 \end{cases}$$

For each $y = (y_1, y_2, \ldots, y_n) \in \Omega$, let $M^y = \bigotimes_{i=1}^{n} M_i^{y_i}$. Then, the bound in (8.3.5) can be simplified further as given in the following theorem. For a proof of the theorem we refer to Gupta and Mukerjee (1989).

<u>Theorem 8.3.2.</u>

$$E_d(\underline{\xi}^x) \geq \frac{\sum_{y \in \Omega^*(x)} w(y)}{\sum_{y \in \Omega^*(x)} w(y) / E_{d_0}(M^y H' B^{-1} \underline{\xi}^x)}$$

where

$$w(y) = \left[\prod_{i:\, x_i = 0,\, y_i = 0} \left(\frac{m_i^2}{q_i} \right) \right] \left[\prod_{i:\, x_i = 0,\, y_i = 1} \left(\underline{1}_i' B_i^{-1} \underline{1}_i - \frac{m_i^2}{q_i} \right) \right]$$

It is interesting to note that the lower bound in the above theorem is in terms of a harmonic mean of efficiencies with respect to the interaction contrasts in d_0. Consequently, by appropriately choosing d_0, one can ensure high efficiencies with respect to the interaction contrasts of interest in d. If d_0 is balanced in addition to being equireplicate and having OFS, then the lower bound in the above theorem is exactly attained. For details we refer to Gupta and Mukerjee (1989). Although the initial design d_0 is equireplicate with OFS, the resulting design d may not have OFS. Still then, exactly as in Section 6.5, a g-inverse of the intra-block matrix of d can be computed in a relatively simple manner.

All the results in this section naturally extend themselves to designs for multiway elimination of heterogeneity.

8.4. Results on efficiency and admissibility.

In the context of a factorial design, one is primarily interested in individual interactions and this emphasis should be taken into account while searching for efficient designs. Dean and Lewis (1980) demonstrated that a design with high overall efficiency does not necessarily have high efficiency with respect to each interaction. Within a given class of designs, it is, therefore, important to find upper bounds for the interaction efficiencies. Such bounds can be used in assessing the performance of a given design in the class with respect to the individual interactions. Following Dean and Lewis (1983), Lewis and Dean (1984) and Bailey (1985a), we present here an account of the available upper bounds for interaction efficiencies.

Consider the class of connected, equireplicate, proper $m_1 \times m_2 \times \cdots \times m_n$ factorial block designs with OFS and having a common replication number r and a constant block size k. Let d be a design in the class and C be the intrablock matrix of d. Since d has OFS, by Theorem 4.2.8, for each $y \in \Omega$, C has $\alpha(y) = \Pi(m_i - 1)^{y_i}$ eigenvalues, say $\lambda_j^y (1 \le j \le \alpha(y))$, with the corresponding orthonormal eigenvectors belonging to the column space of $P^{y\prime}$. For each $y \in \Omega$, it is also clear from the proof of Theorem 4.2.8 that the $\lambda_j^y (1 \le j \le \alpha(y))$ are the eigenvalues of $P^y C P^{y\prime}$, which is, in fact, the information matrix for $P^y \underline{\tau}$ since d has OFS. The connectedness of d implies that $\lambda_j^y > 0$, $1 \le j \le \alpha(y)$, $y \in \Omega$. Then as in (5.3.3), the (A-)efficiency of d with respect to the interaction F^y is defined as

$$\epsilon(y) = r^{-1}\alpha(y)\{\sum_{j=1}^{\alpha(y)} (\lambda_j^y)^{-1}\}^{-1}, \quad y \in \Omega. \tag{8.4.1}$$

The relation (8.4.1) is, as usual, relative to the corresponding randomized (complete) block experiment with the same number of replications as d and corresponds to the special case of Φ_p-efficiency (see Chapter 6) given by $p = 1$.

For $x \in \Omega$, as in Section 4.5, let d_x be the design formed from d by deleting the ith digit from the treatment labels for all i for which $x_i = 0$. The intrablock matrix of d_x is given by $C_x = \epsilon^x C \epsilon^{x'}$, where ϵ^x is given by (4.5.3) and (4.5.4). Let $v_x = \Pi m_i^{x_i}$, $v = \Pi m_i$.

Lemma 8.4.1. For each $x \in \Omega$, the non-zero eigenvalues of C_x are given by $(v/v_x)\lambda_j^y$, $1 \leq j \leq \alpha(y)$, $y \in \Omega(x)$, where $\Omega(x) = \{y: \ y \in \Omega, y_i \leq x_i, \forall_i\}$.

Proof. As in the proof of Theorem 4.2.8, for each $y \in \Omega$, let $\Lambda^y = Diag(\lambda_1^y, \ldots, \lambda_{\alpha(y)}^y)$ and L^y be a $v \times \alpha(y)$ matrix whose columns represent orthonormal eigenvectors of C corresponding to the λ_j^y, $1 \leq j \leq \alpha(y)$. Then a spectral decomposition of C yields (vide 4.2.9))

$$C = \sum_{y \in \Omega} L^y \Lambda^y L^{y'}. \tag{8.4.2}$$

Now, as noted above, the columns of L^y belong to the column space of $P^{y'}$. It is easy to see that $\epsilon^x P^{y'} = 0$ whenever $y \notin \Omega(x)$. Therefore, $\epsilon^x L^y = 0$ whenever $y \notin \Omega(x)$, and by (8.4.2),

$$C_x = \epsilon^x C \epsilon^{x'} = \sum_{y \in \Omega(x)} \epsilon^x L^y \Lambda^y L^{y'} \epsilon^{x'} = \sum_{y \in \Omega(x)} G_y \{(v/v_x) \Lambda^y\} G_y', \tag{8.4.3}$$

where $G_y = (v_x/v)^{1/2} \epsilon^x L^y$. By (2.2.8), for every $y, z \in \Omega(x)$, $y \neq z$, $L^{y'} L^y = I$ and $L^{y'} L^z = 0$. Hence by (4.5.3), (4.5.4), for every $y, z \in \Omega(x), y \neq z$, $G_y' G_y = I$, $G_y' G_z = 0$. Thus (8.4.3) represents a spectral decomposition of C_x and the lemma is proved.

The design d_x is equireplicate with common replication number $r_x = rv/v_x$. Hence by Lemma 8.4.1 and in analogy with (8.4.1), the overall (A-)efficiency of d_x is given by

$$\epsilon^*(x) = r^{-1}(v_x - 1)\{\sum_{y \in \Omega(x)} \sum_{j=1}^{\alpha(y)} (\lambda_j^y)^{-1}\}^{-1}. \tag{8.4.4}$$

As observed in Dean and Lewis (1983) and Lewis and Dean (1984), by (8.4.1), (8.4.4),

$$(v_x-1)\{\epsilon^*(x)\}^{-1} = \sum_{y\in\Omega(x)} \alpha(y)\{\epsilon(y)\}^{-1}. \tag{8.4.5}$$

It may be noted that (8.4.5) was proved for the class of generalized cyclic designs in an unpublished manuscript by J. A. John. By (8.4.4), lemma 8.4.1 and the arithmetic mean-harmonic mean inequality,

$$\epsilon^*(x) \le tr(C_x)/\{r_x(v_x-1)\}. \tag{8.4.6}$$

It can be seen (cf. Williams and Patterson (1977)) that

$$tr(C_x) \le r_x v_x\{1 - k^{-1}w_x\}, \tag{8.4.7}$$

where

$$w_x = \{v_x\rho_x^2 + (k-v_x\rho_x)(2\rho_x+1)\}/k, \tag{8.4.8}$$

and ρ_x is the greatest integer in k/v_x. By (8.4.6), (8.4.7),

$$\epsilon^*(x) \le v_x(1-k^{-1}w_x)/(v_x-1) = U(x), \tag{8.4.9}$$

say. Evidently, by (8.4.5), (8.4.9)

$$(v_x-1)\{U(x)\}^{-1} \le \sum_{y\in\Omega(x)} \alpha(y)\{\epsilon(y)\}^{-1}. \tag{8.4.10}$$

As noted by Lewis and Dean (1984), the relation (8.4.10), which holds for every design in the class under consideration, can be used to obtain an upper bound for $\epsilon(x)$ for particular values of $\epsilon(y)$, $y \in \Omega(x)$, $y \ne x$. In particular, considering a two-factor interaction, say F_1F_2, (8.4.10) yields

$$(m_1m_2-1)U_{12}^{-1} \le (m_1-1)\epsilon_1^{-1} + (m_2-1)\epsilon_2^{-1} + (m_1-1)(m_2-1)\epsilon_{12}^{-1}, \tag{8.4.11}$$

where, for the sake of notational simplicity, we write U_{12}, ϵ_1, ϵ_2 and ϵ_{12} for $U(1,1,0,\ldots,0)$, $\epsilon(1,0,0,\ldots,0)$, $\epsilon(0,1,0,\ldots,0)$ and $\epsilon(1,1,0,\ldots,0)$ respectively. Of course, $0 < \epsilon_1, \epsilon_2, \epsilon_{12} \le 1$. Lewis and Dean (1984) considered a geometric representation of the feasible region for $(\epsilon_1,\epsilon_2,\epsilon_{12})$ subject to (8.4.11) and discussed in some detail the problem of ensuring a high value of ϵ_{12} for given ϵ_1 and ϵ_2.

Earlier, Dean and Lewis (1983) considered binary, equireplicate, proper two-factor designs which are connected, have OFS and allow one or both main effects to be estimated with full efficiency. They employed relations similar to (8.4.11) for obtaining upper bounds for ϵ_{12} for such designs. Bailey (1985a) derived more general results which yield bounds tighter than those in Dean and Lewis (1983) and Lewis and Dean (1984).

Definition 8.4.1 (Bailey, 1985a). An $m_1 \times m_2 \times \cdots \times m_n$ factorial block design d with a constant block size k is x-regular for some $x \in \Omega$ if each element of the incidence matrix of d_x equals either ρ_x or $\rho_x + 1$, where ρ_x is the greatest integer in k/v_x. The design is x-super-regular if it is y-regular for every $y \in \Omega(x)$.

Theorem 8.4.1 (Bailey, 1985a). For a connected, equireplicate, proper $m_1 \times m_2 \times \cdots \times m_n$ factorial block design with OFS and having a constant block size k,

$$\epsilon(x) \leq [v_x - k^{-1} v_x w_x - \sum_{\substack{y \in \Omega(x) \\ y \neq x}} \alpha(y)\epsilon(y)]/\alpha(x), \quad \forall x \in \Omega, \tag{8.4.12}$$

where w_x is as in (8.4.8). Furthermore, if the design is x-super-regular then

$$\epsilon(x) \leq 1 - \{(-1)^{\Sigma x_i} + k^{-1} \sum_{y \in \Omega(x)} (-1)^{\Sigma(x_i - y_i)} v_y w_y\}/\alpha(x) \tag{8.4.13}$$

Proof. For each $y \in \Omega$, by (8.4.1) and the arithmetic mean-harmonic mean inequlaity,

$$\epsilon(y) \leq \beta(y)/\alpha(y), \tag{8.4.14}$$

where $\beta(y) = \sum_{j=1}^{\alpha(y)} \lambda_j^y / r$. Therefore, for each $x \in \Omega$, it follows from Lemma 8.4.1. that

$$r_x^{-1} tr(C_x) = \sum_{y \in \Omega(x)} \beta(y) = \beta(x) + \Sigma' \beta(y) \geq \beta(x) + \Sigma' \alpha(y)\epsilon(y), \tag{8.4.15}$$

where, as before, $r_x = rv/v_x$, and Σ' denotes summation over $y \in \Omega(x)$, $y \neq x$. By (8.4.7), (8.4.14), (8.4.15),

$$\begin{aligned} \epsilon(x) \leq \beta(x)/\alpha(x) &\leq \{r_x^{-1} tr(C_x) - \Sigma' \alpha(y)\epsilon(y)\}/\alpha(x) \\ &\leq \{v_x - k^{-1} v_x w_x - \Sigma' \alpha(y)\epsilon(y)\}/\alpha(x), \end{aligned}$$

proving (8.4.12).

If the design is x-super-regular then it can be seen that

$$r_y^{-1} tr(C_y) = v_y\{1 - k^{-1} w_y\}, \quad \forall y \in \Omega(x). \tag{8.4.16}$$

Also by Lemma 8.4.1,

$$r_y^{-1} tr(C_y) = \sum_{z \in \Omega(y)} \beta(z), \quad \forall y \in \Omega(x)$$

which implies that

$$\beta(x) = \sum_{y \in \Omega(x)} (-1)^{\Sigma(x_i - y_i)} r_y^{-1} tr(C_y). \tag{8.4.17}$$

By (8.4.14), (8.4.16), (8.4.17),

$$\epsilon(x) \le \beta(x)/\alpha(x) = \sum_{y \in \Omega(x)} (-1)^{\Sigma(x_i - y_i)} v_y\{1 - k^{-1} w_y\}/\alpha(x). \tag{8.4.18}$$

After some algebra, it may be seen that the right-hand members of (8.4.13) and (8.4.18) are equal. This proves (8.4.13).

The bound (8.4.13) is never greater than unity or than bound (8.4.12). Thus this bound is always informative and may be used even when values of $\epsilon(y)$ $(y \in \Omega(x), y \neq x)$ are unknown. However, it is applicable only to x-super-regular designs. For more general factorial designs, bound (8.4.12) should be used. Like many other bounds on efficiencies, this bound is not always informative since it may exceed unity.

A simple application of the arithmetic mean-harmonic mean inequality shows that (8.4.12) implies (8.4.10). Consequently, (8.4.12) provides a tighter bound for $\epsilon(x)$ then (8.4.10). Bailey (1985a) also proved that in the special case of two-factor interactions, the bound (8.4.12) is tighter than those in Dean and Lewis (1983). For the proof and also for some further discussion on the applicability and usefulness of the bounds (8.4.12) and (8.4.13), we refer to the original paper.

Mukerjee and Bose (1989) considered the notion of admissibility in the context of factorial designs. Let $D(b,r)$ be the class of $m_1 \times m_2 \times \cdots \times m_n$ factorial designs in b blocks where each treatment combination is replicated r times. The designs in $D(b,r)$ are not necessarily connected or proper and may not have OFS.

Let d be a design in $D(b,r)$ and for $y \in \Omega$, let λ_j^y, $1 \leq j \leq \alpha(y)$, be the eigenvalues of the information matrix for $P^y\underline{\tau}$ in d. The efficiency of d with respect to the interaction F^y is defined by (8.4.1) when $\lambda_j^y > 0$, $1 \leq j \leq \alpha(y)$. If $\lambda_j^y = 0$ for some j then this efficiency is defined to be zero. Hereafter, we shall write $\epsilon_d(y)$ for $\epsilon(y)$ to emphasize that it relates to the design d.

Definition 8.4.2. A design $d_0 \in D(b,r)$ is admissible in $D(b,r)$ if there exists no design $d_1 \in D(b,r)$ such that

$$\epsilon_{d_0}(y) \leq \epsilon_{d_1}(y), \quad \forall y \in \Omega,$$

with strict inequality for at least one y.

In Chapter 3, we considered designs with the C-matrix having property A. It was seen in Theorem 3.1.1 that such designs are balanced with OFS. The next result demonstrates the admissibility of these designs.

Theorem 8.4.2 (Mukerjee and Bose, 1989). Within the class $D(b,r)$ let there exist a design d_0 which is binary and whose C-matrix has property A. Then d_0 is admissible in $D(b,r)$.

Proof. If possible suppose d_0 is not admissible. Then there exists a design $d_1 \in D(b,r)$ such that

$$\epsilon_{d_0}(y) \leq \epsilon_{d_1}(y), \quad \forall y \in \Omega, \tag{8.4.19}$$

with strict inequality for at least one y.

Let $v = \Pi m_i$ and C_0 and C_1 be the C-matrices of d_0 and d_1 respectively. Since C_0 has property A, as in the proof of Theorem 3.1.1, it can be expressed as

$$C_0 = \sum_{y \in \Omega} \lambda^y M^y, \tag{8.4.20}$$

where the λ^y are scalars and the M^y are defined by (2.2.1). hence by (2.2.7), (2.2.8), for each $y \in \Omega$, the information matrix for $P^y\underline{\tau}$ in d_0 is $\lambda^y I$, where I is the identity matrix of order $\alpha(y)$. Therefore,

$$\epsilon_{d_0}(y) = \lambda^y / r, \quad \forall y \in \Omega. \tag{8.4.21}$$

Let for $y \in \Omega$, A_y be the information matrix for $P^y \underline{\tau}$ in d_1. It can be seen (cf. (4.5.1) for the case of connected designs) that $P^y C_1 P^{y'} - A_y$ is non-negative definite, so that

$$\epsilon_{d_1}(y) \leq \{r\alpha(y)\}^{-1} tr(A_y) \leq \{r\alpha(y)\}^{-1} tr(P^y C_1 P^{y'}), \ \forall y \in \Omega. \tag{8.4.22}$$

Since (8.4.19) is strict for at least one y, it follows from (8.4.19), (8.4.21), (8.4.22) that

$$\sum_{y \in \Omega} \alpha(y)\lambda^y < \sum_{y \in \Omega} tr(P^y C_1 P^{y'}) \tag{8.4.23}$$

But $tr(M^y) = \alpha(y)$, so that by (8.4.20),

$$\sum_{y \in \Omega} \alpha(y)\lambda^y = tr(C_0) = vr - b, \tag{8.4.24}$$

recalling that d_0 is a binary design in $D(b,r)$.

Also, as in (3.1.4), defining $P = (\ldots, P^{y'}, \ldots)'$, where P^y is included in P for each $y \in \Omega$, one obtains by Lemma 3.1.2 that

$$\sum_{y \in \Omega} tr(P^y C_1 P^{y'}) = tr(C_1 P' P) = tr(C_1) \leq vr - b, \tag{8.4.25}$$

since $d_1 \in D(b,r)$. The relations (8.4.24) and (8.4.25) contradict (8.4.23). Hence the result follows.

Remark. The condition that d_0 is binary is essential in proving Theorem 8.4.2. Examples may be given to demonstrate that a design in $D(b,r)$ which is not binary but has a C-matrix with property A may not be admissible in $D(b,r)$.

8.5. Concluding remarks

The applications of the calculus for factorial arrangements to various problems concerning characterization, construction and efficiency in traditional factorial experiments have been considered in this monograph. In the concluding section, we briefly indicate several other applications of the calculus.

Sometimes in a factorial experiment, the levels of one factor represent different qualities of material while the levels of another factor represent different quantities of these qualities. Such a situation was first considered by Fisher (1935) in the context of agricultural experiments where zero, single and double doses of certain fertilizers were applied and the yield was studied. Another interesting example was given by John and Quenouille (1977). The important feature of these experiments that make them different form ordinary factorials is that some of the level combinations, namely those where the quantitative factor is at zero level, are indistinguishable. This calls for substantial modifications of the calculus for factorial arrangements. Recently, Bose and Mukerjee (1987) attempted to develop a mathematical formulation for the problem, derived a characterization for OFS by suitably modifying Theorem 4.2.1, and indicated some methods of construction.

Fletcher and John (1985) considered factorial experiments in changeover designs where each experimental unit receives a number of treatment combinations in succession and residual effects of the treatment combinations have to be taken into account in addition to their direct effects. They derived a characterization for orthogonality between direct and residual effects. Also, Theorem 4.2.1 was employed to obtain a characterization for OFS in terms of structure K. It was shown that the generalized cyclic changeover designs have OFS. In a slightly different context, Sen and Mukerjee (1987) used Theorem 4.2.7 to examine how far some optimality results, due to Cheng and Wu (1980) and Magda (1980), on repeated measurements designs remain robust under the possible presence of interaction between direct and residual effects of treatments.

Mukerjee and Huda (1988) considered the design problem for the estimation of variance components by the method of unweighted squares of means under a multifactor random effects model. Employing the calculus for factorial arrangements they established the optimality of balanced designs.

Techniques similar to the calculus have been applied in some other areas, e.g., in discrete multivariate analysis (Haberman (1974, 1975), Mukerjee (1985)), in ANOVA-type decomposition of square integrable statistics (Bhargava (1980), Efron and Stein (1981), Karlin and Rinott (1982)) and also in randomized response techniques for multiattribute situations (Bourke (1981), Mukerjee (1981b)). These aspects are, however, beyond the scope of the present monograph and for further details and references, we refer to Takemura (1983).

REFERENCES

Aggarwal, K. R. (1974). Some higher class PBIB designs and their application as confounded factorial experiments. *Ann. Inst. Statist. Math.* *26*, 315-323.

Bailey, R. A. (1977). Patterns of confounding in factorial designs. *Biometrika 64*, 597-603.

Bailey, R. A. (1985a). Balance, orthogonality and efficiency factors in factorial designs. *J. R. Statist. Soc. B47*, 453-458.

Bailey, R. A. (1985b). Factorial designs and abelian groups. *Linear Algebra and Appl.* *70*, 349-368.

Bailey, R. A., Gilchrist, F. M. L. and Patterson, H. D. (1977). Identification of effects and confounding patterns in factorial designs. *Biometrika 64*, 347-354.

Bhargava, R. P. (1980). A property of Jackknife estimation of the variance when more than one observation is omitted. *Technical Report 140*, Stanford University, Department of Statistics.

Bose, R. C. (1947). Mathematical theory of the symmetrical factorial design. *Sankhya 8*, 107-166.

Bose, R. C. and Kishen, K. (1940). On the problem of confounding in general symmetrical factorial design. *Sankhya 5*, 21-36.

Bose, R. C. and Shrikhande, S. S. (1960). On the composition of balanced incomplete block designs. *Can. J. Math. 12*, 177-188.

Bose, M. and Mukerjee, R. (1987). Factorial designs for quality-quantity interaction. *Calcutta Statist. Assoc. Bull. 36*, 141-152.

Bourke, P. D. (1981). On the analysis of some multivariate randomized response designs for categorical data. *J. Statist. Plann. Inf. 5*, 165-170.

Box, G. E. P. and Draper, N. R. (1987). *Empirical Model Building and Response Surfaces.* John Wiley: New York.

Chatterjee, K. and Mukerjee, R. (1986). Some search designs for symmetric and asymmetric factorials. *J. Statist. Plann. Inf. 13*, 357-363.

Chatterjee, S. K. (1982). Some recent developments in the theory of asymmetric factorial experiments - a review. *Sankhya A44*, 103-113.

Chauhan, C. K. (1987). Some results on regularity in a single replicate design. *Commun. Statist. - Theor. Meth. 16*, 2263-2268.

Chauhan, C. K. (1988). On partial regularity in single replicate factorial experiments. Utilitas Math. 33, 205-210.

Chauhan, C. K. and Dean, A. M. (1986). Orthogonality of factorial effects. *Ann. Statist. 14,* 743-752.

Cheng, C. S. and Wu, C. F. (1980). Balanced repeated measurements designs. *Ann. Statist. 8,* 1272-1283.

Collings, B. C. (1984). Generating the intrablock and interblock subgroups for confounding in general factorial experiments. *Ann. Statist. 12,* 1500-1509.

Cottor, S. C. (1974). A general method of confounding in symmetric factorial exeriments. *J. R. Statist. Soc. B36,* 267-276.

Cottor, S. C., John, J. A. and Smith, T. M. F. (1973). Multi-factor experiments in non-orthogonal designs. *J. R. Statist. Soc. B35,* 361-367.

Das, M. N. (1960). Fractional replicates as asymmetrical factorial designs. *J. Indian Soc. Agric. Statist. 12,* 159-174.

Das, M. N. (1964). A somewhat alternative approach for construction of symmetrical factorial designs and obtaining maximum number of factors. *Calcutta Statist. Assoc. Bull. 13,* 1-17.

Das, M. N. and Giri, N. C. (1986). *Design and Analysis of Experiments.* Second Edition. Wiley Eastern: New Delhi.

David, H. A. and Wolock, F. W. (1965). Cyclic designs. *Ann. Math. Statist. 36,* 1526-1534.

Dean, A. M. (1978). The analysis of interactions in single replicate generalized cyclic designs. *J. R. Statist. Soc. B40,* 79-84.

Dean, A. M. and John, J. A. (1975). Single replicate factorial experiments in generalized cyclic designs II. Asymmetrical arrangements. *J. R. Statist. Soc. B37,* 72-76.

Dean, A. M. and Lewis, S. M. (1980). A unified theory of generalized cyclic designs. *J. Statist. Plann. Inf. 4,* 13-23.

Dean, A. M. and Lewis, S. M. (1983). Upper bounds for average efficiency factors of two-factor interactions. *J. R. Statist. Soc. B45,* 252-257.

Dean, A. M. and Lewis, S. M. (1984). Disconnected generalized cyclic designs. *Ohio State University Technical Report 298.*

Dean, A. M. and Lewis, S. M. (1986). A note on the connectivity of generalized cyclic designs. *Commun. Statist. - Theor. Meth. 15,* 3429-3433.

Dey, A. (1985). *Orthogonal Fractional Factorial Designs.* Wiley Eastern: New Delhi.

Efron, B. and Stein, C. (1981). The jackknife estimate of variance. *Ann. Statist. 9,* 586-596.

El Mossadeq, A., Kobilinksy, A. and Collombier, D. (1985). Construction d'orthogonaux dans les groupes abeliens finis eet confusions d'effects dans les plans factoriels. *Linear Algebra and Appl. 70,* 303-320.

Federer, W. T. (1955). *Experimental Design: Theory and Application.* The Macmillan Company: New York.

Federer, W. T. (1980). Some recent results in experimental design with a bibliography. *Internat. Statist. Rev. 48,* 357-368.

Federer, W. T. and Zelen, M. (1966). Analysis of multifactor classifications with unequal number of observations. *Biometrics* 22, 526-552.

Fisher, R. A. (1935). *The Design of Experiments.* Oliver and Boyd: London.

Fisher, R. A. and Yates, F. (1963). *Statistical Tables for Biological Agricultural and Medical Research (6th edition).* Oliver and Boyd: Edinburgh.

Fletcher, D. J. and John, J. A. (1985). Change over designs and factorial structure. *J. R. Statist. Soc. B47,* 117-124.

Giovagnoli, A. (1977). On the construction of factorial designs using abelian group theory. *Rend. Sem. Mat. Univ. Padova 58,* 195-206.

Gupta, S. C. (1983a). A basic lemma and the analysis of block and Kronecker product designs. *J. Statist. Plann. Inf. 7,* 407-416.

Gupta, S. C. (1983b). Some new methods for constructing block designs having orthogonal factorial structure. *J. R. Statist. Soc. B45,* 297-307.

Gupta, S. C. (1985). On Kronecker block designs for factorial experiments. *J. Statist. Plann. Inf. 11,* 227-236.

Gupta, S. C. (1986a). Interaction efficiencies in Kronecker block designs. *J. Statist. Plann. Inf. 14,* 275-279.

Gupta, S. C. (1986b). Efficiency consistency in designs. *Commun. Statist. - Theor. Meth. 15,* 1315-1318.

Gupta, S. C. (1986c). Factorial experiments in four-associate class cyclic PBIB designs. *Calcutta Statist. Assoc. Bull. 35,* 17-24.

Gupta, S. C. (1987a). A note on component-wise Kronecker designs. *Metrika 34,* 283-286.

Gupta, S. C. (1987b). On designs for factorial experiments derivable from generalized cyclic designs. *Sankhya B49,* 90-96.

Gupta, S. C. (1987c). Generating generalized cyclic designs with factorial balance. *Commun. Statist. - Theor. Meth. 16,* 1885-1900.

Gupta, S. (1988). The association matrices of extended group divisible scheme. *J. Statist. Plann. Inf.* 20, 115-120.

Gupta, S. and Mukerjee, R. (1989). Efficient non-equireplicate designs obtained by merging of treatments. *Comp. Statist. & Data Analy.* 8, (to appear).

Habermann, S. J. (1974). *The Analysis of Frequency Data.* The University of Chicago Press: Chicago.

Habermann, S. J. (1975). Direct products and linear models for complete factorial tables. *Ann. Statist. 3,* 314-333.

Hinkelmann, K. and Kempthorne, O. (1963). Two classes of group divisible partial diallel crosses. *Biometrika 50,* 281-291

Jarret, R. G. and Hall, W. B. (1978). Generalized cyclic incomplete block designs. *Biometrika 65,* 397-401.

John, J. A. (1966). Cyclic incomplete block designs. *J. R. Statist. Soc. B28,* 345-360.

John, J. A. (1973a). Factorial experiments in cyclic designs. *Ann. Statist. 1,* 188-194.

John, J. A. (1973b). Generalized cyclic designs in factorial experiments. *Biometrika 60,* 55-63.

John, J. A. (1981). Factorial balance and the analysis of designs with factorial structure. *J. Statist. Plann. Inf. 5,* 99-105.

John, J. A. (1987). *Cyclic Designs.* Chapman and Hall: New York.

John, J. A. and Dean, A. M. (1975). Single replicate factorial experiments in generalized cyclic designs I. Symmetrical arrangements. *J. R. Statist. Soc. B37,* 63-71.

John, J. A. and Lewis, S. M. (1983). Factorial experiments in generalized cyclic row-column designs. *J. R. Statist. Soc. B45,* 245-251.

John, J. A. and Quenouille, M. H. (1977). *Experiments: Design and Analysis.* Oliver and Boyd: London.

John, J. A. and Smith, T. M. F. (1972). Two-factor experiments in non-orthogonal designs. *J. R. Statist. Soc. B34,* 401-409.

John, J. A., Wolock, F. W. and David, H. A. (1972). *Cyclic Designs.* National Bur. Standards Appl. Math. Ser. 62.

John, P. W. M. (1971). *Statistical Design and Analysis of Experiments.* The Macmillan Company: New York.

Jones, B. (1980). Combining two component designs to form a row-and- column design. *Appl. Statist. 29,* 334-337.

Karlin, S. and Rinott, Y. (1982). Applications of ANOVA-type decompositions for comparisons of conditional variance statistics including jackknife estimates. *Ann. Statist. 10,* 485-501.

Kempthorne, O. (1952). *The Design and Analysis of Experiments.* John Wiley: New York.

Khatri, C. G. and Rao, C. R. (1968). Solutions to some functional equations and their applications to characterization of probability distributions. *Sankhya A30,* 167-180.

Kiefer, J. C. (1975). Construction and optimality of generalized Youden Designs. *A Survey of Statistical Design and Linear Models* (J. N. Srivastava ed.). North Holland: Amsterdam, 333-353.

Kishen, K. (1958). Recent developments in experimental design. *Presidential Address to the Section of Statistics of the 45th Indian Science Congress, Madras.*

Kishen, K. and Srivastava, J. N. (1959). Mathematical theory of confounding in asymmetrical and symmetrical factorial designs. *J. Indian Soc. Agric. Statist. 21,* 73-110.

Kishen, K. and Tyagi, B. N. (1963). Partially balanced asymmetrical factorial designs. *Contributions to Statistics.* Presented to Professor P. C. Mahalanobis on his 70th birthday. Pergamon Press, New York, 147-158.

Kishen, K. and Tyagi, B. N. (1964). On the construction and analysis of some balanced asymmetric factorial designs. *Calcutta Statist. Assoc. Bull. 13,* 123-149.

Kramer, C. Y. and Bradley, R. A. (1957). Intrablock analysis for factorial in two associate class group divisible designs. *Ann. Math. Statist. 28,* 349-361.

Kshirsagar, A. M. (1966). Balanced factorial designs. *J. R. Statist. Soc. B28,* 559-569.

Kurkjian, B. and Zelen, M. (1962). A calculus for factorial arrangements. *Ann. Math. Statist. 33,* 600-619.

Kurkjian, B. and Zelen, M. (1963). Applications of the calculus for factorial arrangements. I. Block and direct product designs. *Biometrika 50,* 63-73.

Kuwada, M. and Nishii, R. (1988). On the characteristic polynomial of the information matrix of balanced fractional s^m factorial designs of resolution $V_{p,q}$. *J. Statist Plann. Inf. 18,* 101-114.

Lewis, S. M. (1982). Generators for asymmetrical factorial experiments. *J. Statist. Plann. Inf. 6,* 59-64.

Lewis, S. M. (1986). Composite generalized cyclic row-column designs. *Sankhya B48,* 373-379.

Lewis, S. M. and Dean, A. M. (1980). Factorial experiments in resolvable generalized cyclic designs. *B. I. A. S. 7,* 159-167.

Lewis, S. M. and Dean, A. M. (1984). Upper bounds for factorial efficiency factors. *J. R. Statist. Soc. B46,* 273-278.

Lewis, S. M. and Dean, A. M. (1985). A note on efficiency consistent designs. *J. R. Statist. Soc. B47,* 261-262.

Lewis, S. M. and Dean, A. M. and Lewis, P. H. (1983). Single replicate designs for two factor experiments. *J. R. Statist. Soc. B45,* 224-227.

Lewis, S. M. and Tuck, M. G. (1985). Paired comparison designs for factorial experiments. *Appl. Statist. 34,* 227-234.

Magda, C. (1980). Circular balanced repeated measurements designs. *Commun. Statist. - Theor. Meth 9,* 1901-1918.

Mathews, G. B. (1892). *The Theory of Numbers.* Bell: Cambridge.

Mukerjee, R. (1979). Inter-effect orthogonality in factorial experiments. *Calcutta Statist. Assoc. Bull. 28,* 83-108.

Mukerjee, R. (1980). Further results on the analysis of factorial experiments. *Calculutta Statist. Assoc. Bull. 29,* 1-26.

Mukerjee, R. (1981a). Construction of effectwise orthogonal factorial designs. *J. Statist. Plann. Inf. 5,* 221-229.

Mukerjee, R. (1981b). Inference on confidential characters from survey data. *Calcutta Statist. Assoc. Bull. 30,* 77-88.

Mukerjee, R. (1982). Construction of factorial designs with all main effects balanced. *Sankhya B44,* 154-166.

Mukerjee, R. (1984). Applications of some generalizations of Kronecker product in the construction of factorial designs. *J. Indian Soc. Agric. Statist. 36,* 38-46.

Mukerjee, R. (1985). Fraction selection problem in discrete muiltivariate analysis.

Sankhya A47, 350-356.

Mukerjee, R. (1986). Construction of orthogonal factorial designs controlling interaction efficiencies. *Commun. Statist. - Theor. Meth. 15*, 1535-1548.

Mukerjee, R. and Bose, M. (1988a). Estimability consistency and its equivalence with regularity in factorial designs. *Utilitas Math.* 33, 211-216.

Mukerjee, R. and Bose, M. (1988b). Non-equireplicate Kronecker factorial designs. *J. Statist. Plann. Inf.* 19, 261-267.

Mukerjee, R. and Bose, M. (1989). Admissibility in factorial designs. *Commun. Statist.-Theor. Meth.* (to appear).

Mukerjee, R. and Dean, A. M. (1986). On the equivalence of efficiency- consistency and orthogonal factorial structure. *Utilitas Math. 30*, 145-151.

Mukerjee, R. and Huda, S. (1988). Optimal design for the estimation of variance components. *Biometrika* 75, 75-80.

Mukerjee, R. and Sen, M. (1988). Kronecker factorial designs for multiway elimination of heterogeneity. *Ann. Inst. Statist. Math.* 40, 195-210.

Muller, E. R. (1966). Balanced confounding of factorial experiments. *Biometrika 53*, 507-524.

Nair, K. R. and Rao, C. R. (1941). Confounded designs for asymmetrical factorial experiments. *Science and Culture 7*, 313-314.

Nair, K. R. and Rao, C. R. (1942). Confounded designs for $k \times p^m \times q^n \times \cdots$ type factorial experiments. *Science and Culture 7*, 361-362.

Nair, K. R. and Rao, C. R. (1948). Confounding in asymmetric factorial experiments. *J. R. Statist. Soc. B10*, 109-131.

Nelder, J. A. (1965). The analysis of randomized experiments with orthogonal block structure. I. Block structure and the null analysis of variance. *Proc. R. Soc. A283*, 147-162.

Paik, U. B. and Federer, W. T. (1973). Partially balanced designs and properties A and B. *Commun. Statist. - Theor. Meth. 1*, 331-350.

Paik, U. B. and Federer, W. T. (1974). Analysis of non-orthogonal n-way classifications. *Ann. Statist.* 2, 1000-1021.

Patterson, H. D. (1965). The factorial combination of treatments in rotation experiments. *J. Agric. Sci. 65*, 171-182.

Patterson, H. D. (1976). Generation of factorial designs, *J. R. Statist. Soc. B38*, 175-179.

Patterson, H. D. and Bailey, R. A. (1978). Design keys for factorial experiments. *Appl. Statist. 27,* 335-343.

Pearce, S. C. (1971). Precision in block experiments. *Biometrika 58,* 161-167.

Puri, P. D. and Nigam, A. K. (1975). A note on efficiency balanced designs. *Sankhya B37,* 457-460.

Puri, P. D. and Nigam, A. K. (1976). Balanced factorial experiments I. *Commun. Statist. - Theor. Meth. 5,* 599-619.

Puri, P. D. and Nigam, A. K. (1978). Balanced factorial experiments II. *Commun. Statist. - Theor. Meth. 7,* 591-605.

Puri, P. D. and Nigam, A. K. (1983). Merging of treatments in block designs. *Sankhya B45,* 50-59.

Raghavarao, D. (1971). *Constructions and Combinatorial Problems in Design of Experiments.* John Wiley: New York.

Raktoe, B. L. (1969). Combining elements from distinct finite fields in mixed factorials. *Ann. Math. Statist. 40,* 498-504.

Raktoe, B. L. (1970). Generalized combining of elements from finite fields. *Ann. Math. Statist. 41,* 1763-1767.

Raktoe, B. L., Hedayat, A. and Federer, W. T. (1980). *Factorial Designs.* John Wiley: New York.

Raktoe, B. L., Rayner, A. A. and Chalton, D. O. (1978). On construction of confounded mixed factorial and lattice designs. *Austral. J. Statist. 20,* 209-218.

Rao, C. R. (1956). A general class of quasi-factorial and related designs. *Sankhya 17,* 165-174.

Rao, C. R. (1973a). *Linear Statistical Inference and its Applications.* Second Edition. John Wiley: New York.

Rao, C. R. (1973b). Some combinatorial problems of arrays and applications to design of experiments. *A Survey of Combinatorial Theory,* J. N. Srivastava ed. North Holland: Amsterdam, 349-359.

Rao, C. R. and Mitra, S. K. (1971). *Generalized Inverse of Matrices and its Applications.* John Wiley: New York.

Roberts, A. W. and Verlag, D. E. (1973) *Convex Functions.* Academic Press: New York.

Sardana, M. G. and Das, M. N. (1965). On the construction and analysis of some confounded asymmetrical factorial designs. *Biometrics 21*, 948-956.

Sen, M. and Mukerjee, R. (1987). Optimal repeated measurements designs under interaction. *J. Statist. Plann. Inf. 17*, 81-91.

Shah, B. V. (1958). On balancing in factorial experiments. *Ann. Math. Statist. 29*, 766-779.

Shah, B. V. (1960a). Balanced factorial experiments. *Ann. Math. Statist. 31*, 502-514.

Shah, B. V. (1960b). On a 5×2^2 factorial design. *Biometrics 16*, 115-118.

Sihota, S. S. and Banerjee, K. S. (1981). On the algebraic structures in the construction of confounding plans in mixed factorial designs on the lines of White and Hultquist. *J. Amer. Statist. Assoc. 76*, 996-1001.

Smith, H. J. S. (1861). On systems of linear indeterminate equations and congruences. *Proc. R. Soc. London 11*, 87-89.

Srivastava, J. N. (1978). A review of some recent work on discrete optimal factorial designs for statisticians and experimenters. *Developments in Statistics, Vol. 1*, 267-329. Academic Press: New York.

Srivastava, J. (1987). On the inadequancy of customary orthogonal arrays in quality control in general scientific experimentation and the need of probing designs of higher revealing power. *Commun. Statist. - Theor. Meth. 16*, 2901-2941.

Street, A. P. and Street, D. J. (1987). *Combinatorics of Experimental Design*. Claredon Press: New York.

Street, D. J. (1986). A survey of construction methods for single replicate factorial designs. *Ars Combinatoria 21A*, 147-168.

Suen, C. -Y. and Chakravarti, I. M. (1986). Efficient two-factor balanced designs. *J. R. Statist. Soc. B48*, 107-114.

Takemura, A. (1983). Tensor analysis of anova decomposition. *J. Amer. Statist. Assoc. 78*, 894-900.

Tharthare, S. K. (1965). Generalized right angular designs. *Ann. Math. Statist. 36*, 1535-1553.

Thomson, H. R. and Dick, I. D. (1951). Factorial designs in small blocks derivable from orthogonal Latin squares. *J. R. Statist. Soc. B13*, 126-130.

Tyagi, B. N. (1971). Confounded asymmetric factorial designs. *Biometrics 27*, 229-232.

Vartak, M. N. (1955). On an application of Kronecker product of matrices to statistical designs. *Ann. Math. Statist. 26,* 420-438.

Voss, D. T. (1986). On generalization of the classical method of confounding to asymmetric factorial experiments. *Commun. Statist. - Theor. Meth. 15,* 1299-1314.

Voss, D. T. (1986). First order deletion designs and the construction of efficient nearly - orthogonal factorial designs in small blocks. *J. Amer. Statist. Assoc. 81,* 813-818.

Voss, D. T. (1988). Single-generator generalized cyclic factorial designs as pseudofactor designs. *Ann. Statist.* 16, 1723-1726.

Voss, D. T. and Dean, A. M. (1985). Methods of confounding in single replicate designs. *Tech. Rep. No. 318,* Dept. of Statistics, Ohio State University.

Voss, D. T. and Dean, A. M. (1987). A comparison of classes of single replicate factorial designs. *Ann. Statist. 15,* 376-384.

Voss, D. T. and Dean, A. M. (1988). On confounding in single replicate factorial experiments, *Utilitas Math.* 33, 59-64.

White, D. and Hultquist, R. A. (1965). Construction of confounding plans for mixed factorial designs. *Ann. Math. Statist. 36,* 1256-1271.

Williams, E. R. (1975). A new class of resolvable block designs. *Ph.D. thesis, Edinburgh University.*

Williams, E. R. and Patterson, H. D. (1977). Upper bounds for efficiency factors in block designs. *Austral. J. Statist. 19,* 194-201.

Worthley, R. and Banerjee, K. S. (1974). A general approach to confounding plans in mixed factorial experiments when the number of levels of a factor is any positive integer. *Ann. Statist. 2,* 579-585.

Yates, F. (1937). The design and analysis of factorial experiments. *Imperial Bureau of Soil Science.* Technical Commun. No. 35.

Zelen, M. (1958). Use of group divisible designs for confounded asymmetric factorial experiments. *Ann. Math. Statist. 29,* 22-40.

Zelen, M. and Federer, W. T. (1964). Applications of the calculus for factorial arrangements II. Designs for two-way elimination of heterogeneity. *Ann. Math. Statist. 35,* 658-672.

Zelen, M. and Federer, W. T. (1965). Applications of the calculus for factorial arrangements III. Analysis of factorials with unequal numbers of observations. *Sankhya A* 27, 383-400.

INDEX